Algebra Workbook FOR DUMMIES®
STUDENT EDITION

by Mary Jane Sterling

Other *For Dummies* math titles:

Algebra For Dummies 0-7645-5325-9
Calculus For Dummies 0-7645-2498-4
Calculus Workbook For Dummies 0-7645-8782-x
Geometry For Dummies 0-7645-5324-0
Statistics For Dummies 0-7645-5423-9
Statistics Workbook For Dummies 0-7645-8466-9
TI-89 Graphing Calculator For Dummies 0-7645-8912-1 (also available for TI-83 and TI-84 models)
Trigonometry For Dummies 0-7645-6903-1
Trigonometry Workbook For Dummies 0-7645-8781-1

Wiley Publishing, Inc.

Algebra Workbook For Dummies®, Student Edition
Published by
Wiley Publishing, Inc.
111 River St.
Hoboken, NJ 07030-5774
www.wiley.com

Copyright © 2006 by Wiley Publishing, Inc., Indianapolis, Indiana

Published by Wiley Publishing, Inc., Indianapolis, Indiana

Published simultaneously in Canada

No part of this publication may be reproduced, stored in a retrieval system, or transmitted in any form or by any means, electronic, mechanical, photocopying, recording, scanning, or otherwise, except as permitted under Sections 107 or 108 of the 1976 United States Copyright Act, without either the prior written permission of the Publisher, or authorization through payment of the appropriate per-copy fee to the Copyright Clearance Center, 222 Rosewood Drive, Danvers, MA 01923, 978-750-8400, fax 978-646-8600. Requests to the Publisher for permission should be addressed to the Legal Department, Wiley Publishing, Inc., 10475 Crosspoint Blvd., Indianapolis, IN 46256, 317-572-3447, fax 317-572-4355, or online at http://www.wiley.com/go/permissions.

Trademarks: Wiley, the Wiley Publishing logo, For Dummies, the Dummies Man logo, A Reference for the Rest of Us!, The Dummies Way, Dummies Daily, The Fun and Easy Way, Dummies.com and related trade dress are trademarks or registered trademarks of John Wiley & Sons, Inc. and/or its affiliates in the United States and other countries, and may not be used without written permission. All other trademarks are the property of their respective owners. Wiley Publishing, Inc., is not associated with any product or vendor mentioned in this book.

LIMIT OF LIABILITY/DISCLAIMER OF WARRANTY: THE PUBLISHER AND THE AUTHOR MAKE NO REPRESENTATIONS OR WARRANTIES WITH RESPECT TO THE ACCURACY OR COMPLETENESS OF THE CONTENTS OF THIS WORK AND SPECIFICALLY DISCLAIM ALL WARRANTIES, INCLUDING WITHOUT LIMITATION WARRANTIES OF FITNESS FOR A PARTICULAR PURPOSE. NO WARRANTY MAY BE CREATED OR EXTENDED BY SALES OR PROMOTIONAL MATERIALS. THE ADVICE AND STRATEGIES CONTAINED HEREIN MAY NOT BE SUITABLE FOR EVERY SITUATION. THIS WORK IS SOLD WITH THE UNDERSTANDING THAT THE PUBLISHER IS NOT ENGAGED IN RENDERING LEGAL, ACCOUNTING, OR OTHER PROFESSIONAL SERVICES. IF PROFESSIONAL ASSISTANCE IS REQUIRED, THE SERVICES OF A COMPETENT PROFESSIONAL PERSON SHOULD BE SOUGHT. NEITHER THE PUBLISHER NOR THE AUTHOR SHALL BE LIABLE FOR DAMAGES ARISING HEREFROM. THE FACT THAT AN ORGANIZATION OR WEBSITE IS REFERRED TO IN THIS WORK AS A CITATION AND/OR A POTENTIAL SOURCE OF FURTHER INFORMATION DOES NOT MEAN THAT THE AUTHOR OR THE PUBLISHER ENDORSES THE INFORMATION THE ORGANIZATION OR WEBSITE MAY PROVIDE OR RECOMMENDATIONS IT MAY MAKE. FURTHER, READERS SHOULD BE AWARE THAT INTERNET WEBSITES LISTED IN THIS WORK MAY HAVE CHANGED OR DISAPPEARED BETWEEN WHEN THIS WORK WAS WRITTEN AND WHEN IT IS READ.

For general information on our other products and services, please contact our Customer Care Department within the U.S. at 800-762-2974, outside the U.S. at 317-572-3993, or fax 317-572-4002.

For technical support, please visit www.wiley.com/techsupport.

Wiley also publishes its books in a variety of electronic formats. Some content that appears in print may not be available in electronic books.

Library of Congress Control Number: 2006924432

ISBN-13: 978-0-470-05666-0

ISBN-10: 0-470-05666-5

Manufactured in the United States of America

10 9 8 7 6 5 4 3 2 1

1B/RR/QV/QW/IN

About the Author

Mary Jane Sterling is also the author of *Algebra For Dummies*, *Trigonometry For Dummies*, *Algebra I CliffsStudySolver*, and *Algebra II CliffsStudySolver*. She taught junior high and high school math for many years before beginning her current 25-years-and-counting tenure at Bradley University in Peoria, Illinois. Mary Jane especially enjoys working with future teachers and trying out new technology.

Publisher's Acknowledgments

We're proud of this book; please send us your comments through our Dummies online registration form located at www.dummies.com/register/.

Some of the people who helped bring this book to market include the following:

Acquisitions, Editorial, and Media Development

Project Editor: Tracy Barr

Acquisitions Editor: Kathy Cox

Editorial Program Coordinator: Hanna K. Scott

Editorial Manager: Michelle Hacker

Editorial Supervisor: Carmen Krikorian

Editorial Assistants: Nadine Bell, Erin Calligan, David Lutton

Cartoons: Rich Tennant (www.the5thwave.com)

Composition Services

Project Coordinator: Maridee Ennis

Layout and Graphics: Carrie A. Foster, Denny Hager, Stephanie D. Jumper, Lynsey Osborn

Proofreaders: Laura Albert, David Faust, Jessica Kramer, Carl Pierce, Mildred Rosenzweig

Indexer: Johnna VanHoose

Publishing and Editorial for Consumer Dummies

Diane Graves Steele, Vice President and Publisher, Consumer Dummies

Joyce Pepple, Acquisitions Director, Consumer Dummies

Kristin A. Cocks, Product Development Director, Consumer Dummies

Michael Spring, Vice President and Publisher, Travel

Kelly Regan, Editorial Director, Travel

Publishing for Technology Dummies

Andy Cummings, Vice President and Publisher, Dummies Technology/General User

Composition Services

Gerry Fahey, Vice President of Production Services

Debbie Stailey, Director of Composition Services

Contents at a Glance

Introduction 1

Part I: Getting a Grip on Basic Concepts 5
Chapter 1: Signing On with Signed Numbers 7
Chapter 2: Making Use of Algebraic Properties 17
Chapter 3: Working with Fractions 25
Chapter 4: Discovering Exponents 47
Chapter 5: Taming the Radicals 57
Chapter 6: Keeping It Simple for Algebraic Expressions 67

Part II: Operating and Factoring 75
Chapter 7: Using Special Rules for Multiplying Expressions 77
Chapter 8: Doing Long Division to Simplify Algebraic Expressions 89
Chapter 9: Factoring Algebraic Expressions 101
Chapter 10: Two at a Time with Factoring 107
Chapter 11: Factoring Trinomials and Other Expressions 113

Part III: Stirring Up Solutions 129
Chapter 12: Putting It on the Line: Solving Linear Equations 131
Chapter 13: Solving Quadratic Equations 153
Chapter 14: Yielding to Higher Powers 167
Chapter 15: Solving Radical and Absolute Value Equations 177
Chapter 16: Working with Inequalities 189

Part IV: Applying Your Skills to Solve Story Problems 205
Chapter 17: Figuring Out Formulas 207
Chapter 18: Applying Formulas to Basic Story Problems 221
Chapter 19: Comparing Things in Story Problems 235
Chapter 20: Weighing In on Quality and Quantity Story Problems 245

Part V: The Part of Tens 257
Chapter 21: Ten (or So) Things to Know about Graphing 259
Chapter 22: Ten Common Pitfalls and How to Avoid 'Em 273
Chapter 23: Ten Quick Tips to Make Algebra a Breeze 277

Index 281

Table of Contents

Introduction 1

    About This Book 1
    Conventions Used in This Book 1
    Foolish Assumptions 2
    How This Book Is Organized 2
        Part I: Getting a Grip on Basic Concepts 2
        Part II: Operating and Factoring 3
        Part III: Stirring Up Solutions 3
        Part IV: Applying Your Skills to Solve Story Problems 3
        Part V: The Part of Tens 3
    Icons Used in This Book 4
    Where to Go from Here 4

Part I: Getting a Grip on Basic Concepts 5

Chapter 1: Signing On with Signed Numbers 7

    Comparing Numbers on the Number Line 7
    Absolutely Right — Writing Absolute Value 8
    Adding Signed Numbers 9
    Making a Difference with Signed Numbers 11
    Multiplying Signed Numbers 12
    Dividing Signed Numbers 13
    Answers to Problems on Signed Numbers 15

Chapter 2: Making Use of Algebraic Properties 17

    Using Grouping Symbols 17
    Distributing the Wealth 19
    Associating Correctly 20
    Commuting to Work 21
    Answers to Problems on Algebraic Properties 23

Chapter 3: Working with Fractions 25

    Converting Improper and Mixed Fractions 25
    Finding Equivalent Fractions 27
    Making It Proportional 28
    Finding Common Denominators 30
    Adding Fractions 32
    Subtracting Fractions 34
    Multiplying Fractions 35
    Dividing Fractions 37
    Changing Fractions to Decimals and Vice Versa 39
    Answers to Problems on Fractions 41

Chapter 4: Discovering Exponents 47

    Multiplying Numbers with Exponents 47
    Dividing Numbers with Exponents 48
    Raising Powers to Powers 49

Using Negative Exponents 51
Writing Numbers with Scientific Notation 52
Answers to Problems on Discovering Exponents 54

Chapter 5: Taming the Radicals 57

Simplifying Radical Expressions 57
Rationalizing Fractions 59
Changing Radicals to Exponents 60
Using Fractional Exponents 61
Simplifying Expressions with Exponents 62
Answers to Problems on Radicals 64

Chapter 6: Keeping It Simple for Algebraic Expressions 67

Adding and Subtracting Like Terms 67
Multiplying and Dividing Algebraic Factors 68
Using Order of Operation 69
Evaluating Expressions with Order of Operations 71
Answers to Problems on Algebraic Expressions 73

Part II: Operating and Factoring 75

Chapter 7: Using Special Rules for Multiplying Expressions 77

Distributing One Factor over Many 77
Getting FOILed Again 78
Squaring Binomials 80
Multiplying the Sum and Difference of the Same Two Terms 81
Cubing Binomials 82
Creating the Sum and Difference of Cubes 83
Raising Binomials to Higher Powers 84
Answers to Problems on Multiplying Expressions 86

Chapter 8: Doing Long Division to Simplify Algebraic Expressions 89

Dividing by a Monomial 89
Dividing by a Binomial 91
Dividing by Other Polynomials 94
Trying Synthetic Division 95
Answers to Problems on Division 97

Chapter 9: Factoring Algebraic Expressions 101

Looking at Prime Factorizations 101
Factoring Out the Greatest Common Factor 102
Reducing Algebraic Fractions 103
Answers to Problems on Factoring Expressions 105

Chapter 10: Two at a Time with Factoring 107

Factoring the Difference of Squares 107
Factoring Differences and Sums of Cubes 108
Factoring in More Than One Way 109
Answers to Problems on Factoring 111

Chapter 11: Factoring Trinomials and Other Expressions 113

Finding the Greatest Common Factor (GCF) 113
"Un"wrapping the FOIL 115
Factoring Trinomials in More Than One Way 118

Factoring by Grouping 120
Putting All the Factoring Together 122
Answers to Problems on Factoring Trinomials and Other Expressions 125

Part III: Stirring Up Solutions 129

Chapter 12: Putting It on the Line: Solving Linear Equations 131

Using the Addition/Subtraction Property 131
Using the Multiplication/Division Property 133
Putting Several Operations Together 134
Solving Linear Equations with Grouping Symbols 136
Working It Out with Fractions 138
Solving Proportions 141
Working with Formulas 143
Answers to Problems on Solving Linear Equations 145

Chapter 13 : Solving Quadratic Equations 153

Using the Square Root Rule 153
Solving by Factoring 154
Using the Quadratic Formula 157
Estimating Answers 160
Dealing with Impossible Answers 162
Answers to Problems on Solving Quadratic Equations 163

Chapter 14: Yielding to Higher Powers 167

Determining the Number of Possible Roots 167
Applying the Rational Root Theorem 168
Using the Factor/Root Theorem 170
Solving By Factoring 171
Solving Powers That Are Quadratic-Like 172
Answers to Problems on Solving Higher Power Equations 174

Chapter 15 : Solving Radical and Absolute Value Equations 177

Solving Radical Equations by Squaring Once 177
Squaring Radical Equations Twice 180
Solving Absolute Value Equations 182
Answers to Problems on Radical and Absolute Value Equations 184

Chapter 16: Working with Inequalities 189

Using the Rules to Work on Inequality Statements 189
Solving Linear Inequalities 190
Solving Quadratic Inequalities 192
Solving Other Inequalities 193
Solving Absolute Value Inequalities 195
Solving Inequalities by Sections 197
Answers to Problems on Working with Inequalities 198

Part IV: Applying Your Skills to Solve Story Problems 205

Chapter 17: Figuring Out Formulas 207

Applying the Pythagorean Theorem 207
Getting Around Using Perimeter Formulas 209

Squaring Off with Area Formulas 210
Working with Volume Formulas 211
Distancing Yourself with the Distance Formula 213
Getting Interested in Using Percent 214
Answers to Problems on Using Formulas 216

Chapter 18 : Applying Formulas to Basic Story Problems 221

Deciphering Perimeter, Area, and Volume 221
Using Geometry to Solve Story Problems 224
Going the Distance with Story Problems 226
Answers to Problems on Using Formulas in Story Problems 229

Chapter 19: Comparing Things in Story Problems 235

Answering Age Problems 235
Tackling Consecutive Integer Problems 237
Working Together on Work Problems 240
Answers to Problems on Comparing Things in Story Problems 242

Chapter 20: Weighing In on Quality and Quantity Story Problems 245

Achieving the Right Blend with Mixtures Problems 245
Finding the Correct Solution One Hundred Percent of the Time 248
Watching Your Pennies with Money Problems 250
Answers to Problems on Weighing Quality and Quantity 252

Part V: The Part of Tens 257

Chapter 21: Ten (or So) Things to Know about Graphing 259

Thickening the Plot with Points 259
Sectioning Off by Quadrants 260
Plotting Points for Lines 261
Graphing Lines with Intercepts 262
Sliding Down Slopes 262
Graphing with the Slope-Intercept Form 263
Changing to the Slope-Intercept Form 264
Lining Up Parallel and Perpendicular Lines 265
Finding Distances Between Points 266
Plotting Parabolas 267
Taking on Intercepts of Parabolas 268
Graphing with Transformations 269

Chapter 22: Ten Common Pitfalls and How to Avoid 'Em 273

Squaring a Power 273
Squaring a Binomial 273
Ordering Around Operations 274
Becoming Radical 274
Distributing a Negative 275
Fracturing Fractions 275
Reducing Fractions 275
Using Negative Exponents 276
Determining Which Is Smaller 276
Reversing the Sense 276

Chapter 23: Ten Quick Tips to Make Algebra a Breeze 277
    Flipping Proportions 277
    Multiplying Through to Get Rid of Fractions 277
    Zooming In on the Zero 278
    Finding a Common Denominator 278
    Dividing by 3 or 9 278
    Dividing by 2, 4, or 8 279
    Commuting Back and Forth 279
    Lining Up with Symmetry 280
    Making Radicals Less Rad, Baby 280
    Eying Up the Polynomial Function 280

Index *281*

Introduction

You now have it in your hands: A workbook that gives you the opportunity to show what you can do in algebra. No, don't panic. You're not going to be doing these problems alone. As you proceed through *Algebra Workbook For Dummies,* you'll see plenty of road signs that clearly mark the way. You'll find plenty of explanations, examples, and other bits of info to make this journey as smooth an experience as possible.

Mathematics is a subject that has to be *handled*. You can read English literature and understand it without having to actually write it. You can read about biological phenomena and understand them, too, without taking part in an experiment. Mathematics is different. You really do have to do it, practice it, play with it, and use it for the mathematics to become a part of your knowledge and skills. And what better way to get your fingers wet than by jumping into this workbook? Remember only practice, practice, and some more practice can help you master algebra! Have at it!

About This Book

I organize *Algebra Workbook For Dummies,* Student Edition, very much like the way I organized *Algebra For Dummies* (Wiley) that you may already have — from the basic concepts and properties to the more complex.

One nice thing about this workbook and other *For Dummies* books is that you don't have to start at the beginning and work your way, step-by-step, from beginning to end. You can start wherever you want, and you don't have to follow any rules about backtracking. If you get stuck on something because you can't remember a certain process, then just put a sticky note where you're stuck, ask your algebra teacher for help, or look up the information in *Algebra For Dummies* or some other helpful book, and then return to the sticky note when you straighten it out.

Of course, you do need the basic algebra concepts to start anywhere in this workbook, but, after you have those down pat, you can pick and choose where you want to work. You can jump in wherever you want and work from there.

Conventions Used in This Book

I use the following conventions in this book to make everything consistent and easy to understand as you solve the plethora of practice problems:

- New terms appear in *italic* and are closely followed by a clear definition.
- I **bold** the answers to the example and the practice questions for easy identification. However, I don't bold the following punctuation to prevent any confusion with periods and decimal points that could be considered part of the answer.
- Algebra uses a lot of letters to represent numbers. In general, I use letters at the beginning of the alphabet (a, b, c, k) to represent *constants* — numbers that don't change all the time but may be special to a particular situation. The letters at the end of the alphabet usually represent *variables* — what you're solving for. I use the most commonly used letters (x and y) for variables. And all constants and variables are *italicized*.

- I use the corresponding symbols to represent the math operations of addition, subtraction, multiplication, and division: +, –, ×, and ÷. But keep the following special rules in mind when using them in algebra and in this book:

    - Subtraction (–) is an operation, but that symbol also represents *opposite of, minus,* and *negative*. When you get to the different situations, you can figure out how the interpretations call for different wording.

    - Multiplication (×) is usually indicated with a dot (for example $3 \cdot 4 = 12$) or parenthesis ( ) in algebra. In this book, I use parentheses most often, but you may occasionally see a × symbol. To avoid any confusion, I don't use the dot in this book; however, remember that when doing algebra problems at school, you'll probably encounter the dot more instead of the ×. Don't confuse the × symbol with the italicized variable, *x*.

    - Division (÷) is sometimes indicated with a slash (/) or fraction line.

Foolish Assumptions

I don't go into great depth when explaining the theories and rules behind each problem. I assume that you have some other book, like my *Algebra For Dummies* (Wiley) for more in-depth reference, if necessary. This workbook provides you with practice problems so you can check to see if you *really* do understand a particular concept or process.

When writing this book, I assume the following about you, my dear reader:

- You already have a decent background in basic algebra concepts and want an opportunity to practice those skills.
- You took or currently are taking Algebra I, but you need to brush up on certain areas.
- Your son, daughter, grandson, granddaughter, niece, nephew, or special someone is taking Algebra I. You haven't looked at an equation for years and you want to help him or her.
- You love math, and your idea of a good time is solving equations on a rainy afternoon.

How This Book Is Organized

Like all books in the *For Dummies* series, this book is divided into parts. This organization allows you to pinpoint where you want to start or where you need to revisit. Each part covers a general area of study or type of concept. The important algebra topics divide up nicely into the following five parts.

Part 1: Getting a Grip on Basic Concepts

This first part starts at the beginning of some algebra topics, but it doesn't start at the beginning of arithmetic or cover many pre-algebra topics. This beginning includes working with signed numbers and their operations. It also includes the ever-loving fractions and what you need to know to add or multiply them. You have to understand fractions, in general, to be able to work with fractions and algebraic terms. This part also focuses on exponents, numbers, and variables and how they combine — or don't combine. I follow exponents with radicals — not the hippies from the 1960s, but those operations that can be represented with fractional exponents. They go together! Lastly, the basics include combining terms that are alike enough and have enough in common.

Part II: Operating and Factoring

Algebra is a stepping-stone to higher mathematics. In fact, you really can't do much advanced mathematics without algebra. There's the cryptic language in the form of letters and operation symbols, and then there are the operations. You need to acquaint yourself with these symbols and operations in order to move on to other algebraic processes such as solving equations and graphing. This part describes and refines the operations of addition, subtraction, multiplication, and division. It also coordinates the operations in terms of having unknown or variable terms and factors. The operations are the same as with numbers; they just look different and have different types of results.

The factoring part is *big*. True, factoring is really just rediscovering what was, at one time, something that got multiplied out. But that — the *what was multiplied out* — is what helps figure out how to factor. You can think of factoring as being the first step in a puzzle or challenge. Get good at factoring, and the answer comes much more easily.

Part III: Stirring Up Solutions

Discovering a solution to an equation is usually everyone's favorite part of algebra. Give me an equation, and I can figure out what makes it work. Sometimes you can just look at an equation, and the solution pops right out at you. For instance, doesn't the equation $x + 1 = 7$ just cry out that $x = 6$? Sometimes you just *think* you know what it is, but beware that there may be more to it than meets the eye. This part discusses the different types of equations and inequalities in terms of their similarities and how to handle those similarities.

Part IV: Applying Your Skills to Solve Story Problems

If you've mastered all the techniques needed to solve the different types of equations, you can focus on writing some equations, and put those skills to use. The applications of algebra come in the form of standard formulas for area, temperature, distance, and many more forms. The applications also take the form of word problems that need to be translated into equation form so you can solve them.

The story problems are divided into many different types, and each type has a specific way to handle it in order to solve it.

Part V: The Part of Tens

Like every *For Dummies* book, the Part of Tens chapters offer you some quick tips. This part has two completely different lists. You can call them the "how to" and "how not to" lists.

The "how to" list includes tricks of the trade that I pull together from several areas. These tips can save you time and energy when dealing with different situations in algebra.

The "how not to" list contains some of the more frequently occurring errors in algebra. Oh, yes, people have plenty of opportunities for errors when dealing with algebraic terms and processes, but some stand out above others. Maybe these common errors are centered deep within the human brain — they fool people again and again for some reason. In any case, look them over to avoid these pitfalls.

Icons Used in This Book

A *For Dummies* book includes icons that help you find and fathom key ideas and information. However with this workbook, the entire book is chock-full of important nuggets of information. As a result, I only highlight the crème-de-la-crème information with icons. They stand out to get your attention — you can't ignore these cute little graphics in the left margin. I hope you find them to be timely and helpful.

You find Example icons everywhere in this book. Before you attempt the problems, look over an example or two, which can help you start. These examples cover all the techniques needed to do the practice problems, but, sometimes, you have to look at the solutions at the end of the chapter for the nitty-gritty details.

These Tip icons are little hints or suggestions to help ease your way through the problems. They appear where some recognized complication occurs with a possible resolution. I can save you time and energy, and cut down on the frustration level with them.

This icon highlights extremely important rules or processes that you should squirrel away in your brain for quick recall later.

Although this icon isn't in red, it does call attention to a particularly troublesome point. When I use this icon, I identify the tricky elements and give you *fair warning*.

Where to Go from Here

Now you're ready to start. You're all packed and ready to go. It's time to take this excursion in algebra.

I hope you packed your enthusiasm and sense of adventure. Yes, this workbook is a grand adventure, just waiting for you to jump. I also recommend a guidebook to help you with the trouble spots. One such guide is my book, *Algebra For Dummies* (Wiley), which is a companion to this book. Furthermore it also mirrors most of the different topics. You can use it to fill in the gaps.

I recommend that you pack a pencil with an eraser. It's the teacher and mathematician in me who realizes that mistakes can be made, and they erase easier when in pencil. That scratched-out blobby stuff bothers me.

Where do you start? No, you don't have to stick with the rest of the tour group. You can venture out on your own, taking your own path, making your own plans, jumping from chapter to chapter. You can do what you want. But, then, you can always stay with the security and grand plan and start with the first chapter and carefully proceed. You don't want to miss anything important.

Part I
Getting a Grip on Basic Concepts

In This Part . . .

Every subject has basic building blocks. The building blocks for algebra actually start back with your first 1 + 1, but this part doesn't go back that far. The building blocks needed to do algebra effectively include important concepts, such as handling negative signs on numbers and putting fractions in exponents. Knowing how to use these elements is essential to the whole subject. Use the chapters in this part as refreshers — or move on if you already have a great foundation.

Chapter 1

Signing On with Signed Numbers

In This Chapter
- Using the number line
- Trying absolute value
- Operating on signed numbers: adding, subtracting, multiplying, and dividing

In this chapter you practice the operations on signed numbers and figure out how to make them behave the way you want them to. (Just tell your mother she can't use this chapter on your little brother to make him behave.) The properties are very helpful in making math expressions easier to read and to handle when solving equations in algebra.

Comparing Numbers on the Number Line

You may think that identifying that 16 is bigger than 10 is an easy concept. But what about –16 and –10? Which is bigger?

The easiest way to compare numbers and to tell which is bigger or has a greater value is to find their position on the number line. The number line goes from negatives on the left to positives on the right (see Figure 1-1). Whichever number is farther to the right has the greater value — it's bigger.

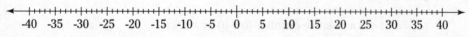

Figure 1-1: A number line.

Q. Using the number line in Figure 1-1, determine which is larger, –16 or –10.

A. **–10.** The number –10 is to the right of –16, so it's the bigger of the two numbers. You write that as –10 > –16 (read this as "negative 10 is greater than negative 16"). Or you can write it as –16 < –10 (negative 16 is less than negative 10).

Q. Which is larger, –.0023 or –.023?

A. **–.0023.** The number –.0023 is to the right of –.023 so it's larger.

8 Part I: Getting a Grip on Basic Concepts

1. Which is larger, –2 or –8?

Solve It

2. Which has the greater value, 0 or –1?

Solve It

3. Which is bigger, –.003 or –.03?

Solve It

4. Which is larger, –⅙ or –⅔?

Solve It

Absolutely Right — Writing Absolute Value

The *absolute value* of a number, written $|a|$, is an operation that evaluates whatever is between the vertical bars and then outputs a positive number. Another way of looking at this operation is it can tell you how far a number is from 0 on the number line — with no reference to which side.

The absolute value of a

$|a| = a$, if a is a positive number ($a > 0$) or if $a = 0$

$|a| = -a$, if a is a negative number ($a < 0$). Read this as "The absolute value of a is equal to the *opposite* of a."

Q. $|4| =$

A. 4

Q. $|-3| =$

A. 3

5. $|8| =$

Solve It

6. $|-6| =$

Solve It

7. $-|-6| =$

Solve It

8. $-|8| =$

Solve It

Adding Signed Numbers

Adding signed numbers involves two different rules.

- You use one when the signs of the two numbers are the same — both positive or both negative.
- You use the other when the two numbers' signs are different.

After you determine whether the signs are the same or different, then you use the absolute values of the numbers.

To add signed numbers

If the signs are the same, add the absolute values of the two numbers together, and let their common sign be the sign of the answer.

$(+a) + (+b) = +(a + b)$

$(-a) + (-b) = -(a + b)$

If the signs are different, then find the difference between the absolute values of the two numbers (subtract the smaller absolute value from the larger), and let the answer have the sign of the number with the larger absolute value. Assume that $|a| > |b|$.

$(+a) + (-b) = +(a - b)$

$(-a) + (+b) = -(a - b)$

10 Part I: Getting a Grip on Basic Concepts

Q. $(-6) + (-4) = -(6 + 4) =$
A. -10

Q. $(+8) + (-15) = -(15 - 8) =$
A. -7

9. $4 + (-3) =$
Solve It

10. $5 + (-11) =$
Solve It

11. $(-18) + (-5) =$
Solve It

12. $47 + (-33) =$
Solve It

13. $(-3) + 5 + (-2) =$
Solve It

14. $(-4) + (-6) + (-10) =$
Solve It

15. 5 + (–18) + (10) =

Solve It

16. (–4) + 4 + (–5) + 5 + (–6) =

Solve It

Making a Difference with Signed Numbers

You really don't use a new set of rules for subtracting signed numbers. You just change the subtraction problem to an addition problem and use the rules for addition of signed numbers. To ensure that your answer to this new addition problem is the answer to the subtraction problem, you not only change the operation from subtraction to addition, but you also change the sign of the second number — the one that's being subtracted.

To subtract two signed numbers

$a - (+b) = a + (-b)$

$a - (-b) = a + (+b)$

Q. (–8) – (–5)

A. –3

Q. What's the average annual rainfall for Scottsdale, Arizona?

A. **7.05 inches.** Huh? Where'd that come from, you're wondering? I just wanted to keep you on your toes! Keep this information in mind the next time you plan a trip to Arizona.

17. 5 – (–2) =

Solve It

18. –6 – (–8) =

Solve It

19. 4 – 87 =

Solve It

20. 0 – (–15) =

Solve It

21. 2.4 – (–6.8) =

Solve It

22. –15 – (–11) =

Solve It

Multiplying Signed Numbers

When you multiply two expressions with the same sign, the product is positive; when the two expressions have different signs, the product is negative. If you're multiplying more than two factors together, just count the number of negative signs in the problem. If the number of negative signs is an even number, then the answer is positive. If the number of negative signs is odd, then the answer is negative.

The product of two signed numbers

$(+)(+) = +$

$(-)(-) = +$

$(+)(-) = -$

$(-)(+) = -$

The product of more than two signed numbers

$(+)(+)(+)(-)(-)(-)(-)$ has a *positive* answer for an *even* number of negative factors.

$(+)(+)(+)(-)(-)(-)$ has a *negative* answer for an *odd* number of negative factors.

Q. (–2)(–3) =

A. +6

Q. (–2)(+3) =

A. –6

Chapter 1: Signing On with Signed Numbers 13

23. (–6)(3) =

Solve It

24. (14)(–1) =

Solve It

25. (–6)(–3) =

Solve It

26. (6)(–3)(4)(–2) =

Solve It

27. (–1)(–1)(–1)(–1)(–1)(2) =

Solve It

28. (–10)(2)(3)(1)(–1) =

Solve It

Dividing Signed Numbers

The rules for dividing signed numbers are exactly the same as those for multiplying signed numbers — as far as the sign goes. (See "Multiplying Signed Numbers" earlier in this chapter.) They differ though because you have to divide, of course.

When dividing signed numbers, just count the number of negative signs that are in the problem — in the numerator, the denominator, and perhaps in front of the problem. If you have an even number of negative signs, the answer is positive. If you have an odd number of negative signs, the answer is negative. This rule works only when factors are multiplied and divided, not added or subtracted.

Q. $^{-36}/_{-9} =$

A. +4

Q. $\dfrac{-(-3)(-12)}{4} =$

A. −9

29. $^{-22}/_{-11} =$

Solve It

30. $^{24}/_{-3} =$

Solve It

31. $\dfrac{-3(-4)}{-2} =$

Solve It

32. $\dfrac{(-5)(2)(3)}{-1} =$

Solve It

33. $\dfrac{(-2)(-3)(-4)}{(-1)(-6)} =$

Solve It

34. $^{(-1)}/_{(-1)} =$

Solve It

Answers to Problems on Signed Numbers

This section provides the answers (in bold) to the practice problems in this chapter.

1 Which is larger, –2 or –8? **–2 is larger**. The following number line shows that the number –2 is to the right of –8. So –2 is bigger than –8 (or –2 > –8).

2 Which has the greater value, 0 or –1? **0 is greater**. The number 0 is to the right of –1. So 0 has a greater value than –1 (or 0 > –1).

3 Which is bigger, –.003 or –.03? **–.003 is bigger**. The following number line shows that the number –.003 is to the right of –.03, which means –.003 is bigger than –.03 (or –.003 > –.03).

4 Which is larger, –⅙ or –⅔? **–⅙ is larger**. The number –⅔ = –⁴⁄₆, and –⁴⁄₆ is to the left of –⅙ on the following number line. So –⅙ is larger than –⅔ (or –⅙ > –⅔).

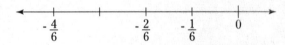

5 $|8| = $ **8** because 8 > 0.

6 $|-6| = $ **6** because –6 < 0 and 6 is the opposite of –6.

7 $-|-6| = $ **–6** because $|-6| = 6$ as in the previous problem.

8 $-|8| = $ **–8** because $|8| = 8$.

9 4 + (–3) = **1** because 4 is the greater absolute value.
4 + (–3) = +(4 – 3) = 1

10 5 + (–11) = **–6** because –11 has the greater absolute value of 11.
5 + (–11) = –(11 – 5) = –6

11 (–18) + (–5) = **–23** because both of the numbers have negative signs; when the signs are the same, find the sum of their absolute values. (–18) + (–5) = –(18 + 5) = –23

12 47 + (–33) = **14** because 47 has the greater absolute value. 47 + (–33) = +(47 – 33) = 14

13 (–3) + 5 + (–2) = **0**
(–3) + 5 + (–2) = ((–3) + 5) + (–2) = (2) + (–2) = 0

14 (–4) + (–6) + (–10) = **–20**
(–4) + (–6) + (–10) = –(4 + 6) + (–10) = (–10) + (–10) = –(10 + 10) = –20

15 $5 + (-18) + (10) = -3$

$5 + (-18) + (10) = -(18 - 5) + 10 = -(13) + 10 = -(13 - 10) = -3$

Or you may prefer to add the two numbers with the same sign, first, like this: $5 + (-18) + (10) = (5 + 10) + (-18) = 15 + (-18) = -(18 - 15) = -3$

16 $(-4) + 4 + (-5) + 5 + (-6) = -6$

$(-4) + 4 + (-5) + 5 + (-6) = \big((-4) + 4\big) + \big((-5) + 5\big) + (-6) = 0 + 0 + (-6) = -6$

17 $5 - (-2) = 7$

$5 - (-2) = 5 + (+2) = 7$

18 $-6 - (-8) = 2$

$-6 - (-8) = -6 + (+8) = 8 - 6 = 2$

19 $4 - 87 = -83$

$4 - 87 = -(87 - 4) = -83$

20 $0 - (-15) = 15$

$0 - (-15) = 0 + 15 = 15$

21 $2.4 - (-6.8) = 9.2$

$2.4 - (-6.8) = 2.4 + 6.8 = 9.2$

22 $-15 - (-11) = -4$

$-15 - (-11) = -15 + 11 = -(15 - 11) = -4$

23 $(-6)(3) = -18$ because the multiplication problem has one negative, and one is an odd number.

24 $(14)(-1) = -14$ because the multiplication problem has one negative, and one is an odd number.

25 $(-6)(-3) = 18$ because the multiplication problem has two negatives, and two is even.

26 $(6)(-3)(4)(-2) = 144$ because the multiplication problem has two negatives.

27 $(-1)(-1)(-1)(-1)(-1)(2) = -2$ because the multiplication problem has five negatives.

28 $(-10)(2)(3)(1)(-1) = 60$ because the multiplication problem has two negatives.

29 $\frac{-22}{-11} = 2$ because the division problem has two negatives, and two is an even number.

30 $\frac{24}{-3} = -8$ because the division problem has one negative, and one is odd.

31 $\frac{-3(-4)}{-2} = -6$ because three negatives result in a negative.

32 $\frac{(-5)(2)(3)}{-1} = 30$ because the division problem has two negatives.

33 $\frac{(-2)(-3)(-4)}{(-1)(-6)} = -4$ because the division problem has five negatives.

34 $\frac{(-1)}{(-1)} = 1$ because the division problem has two negatives.

Chapter 2
Making Use of Algebraic Properties

In This Chapter
- Understanding different types of grouping symbols
- Distributing over addition and multiplication
- Utilizing the associative and commutative rules

Algebra has rules for everything including a sort of shorthand notation to save time and space. This notation also cuts down on misinterpretation — it's very specific and universally known. (I give the rules for doing operations like addition, subtraction, multiplication, and division in Chapter 1.) In this chapter, you see the rules for using grouping symbols and rearranging terms.

Using Grouping Symbols

The most commonly used *grouping symbols* in algebra are (in order of most common):

- Parentheses ( )
- Brackets [ ]
- Braces { }
- Absolute value symbols | |
- Radicals $\sqrt{}$
- Fraction lines /

What's understood about grouping symbols is that whatever is inside them (or under or over, in the case of the fraction line) has to be computed first, before you can use that result to solve the problem. Or, if what's inside isn't or can't be simplified into one term, then anything outside the grouping symbol that multiplies one of the terms has to multiply them all — that's the *distributive property,* found in the very next section.

Q. $16 - (4 + 2) =$

A. **10**. Add the 4 and 2, and then subtract the result from the 16: $16 - (4 + 2) = 16 - 6 = 10$

Q. Simply $2[6 - (3 - 7)]$.

A. **20**. First subtract the 7 from the 3, and then subtract the –4 from the 6 by changing it to an addition problem. You can then multiply the 2 by the 10: $2[6 - (3 - 7)] = 2[6 - (-4)] = 2[6 + 4] = 2[10] = 20$

Part I: Getting a Grip on Basic Concepts

Q. $1 - |-8 + 19| + 3\{4 + 2\} =$

A. 8. Combine what's in the absolute value and braces first, before combining the results:

$$1 - |-8 + 19| + 3\{4 + 2\} = 1 - |11| + 3\{6\}$$
$$= 1 - 11 + 18$$
$$= 19 - 11 = 8$$

Q. $\dfrac{32}{30 - 2(3 + 4)} =$

A. 2. You have to complete the work in the denominator first before dividing the 32 by that result:

$$\dfrac{32}{30 - 2(3 + 4)} = \dfrac{32}{30 - 2(7)} = \dfrac{32}{30 - 14} = \dfrac{32}{16} = 2$$

---

1. $3(2 - 5) + 14 =$

Solve It

2. $4[3(6 - 8) + 2(5 + 9)] - 11 =$

Solve It

---

3. $5\{8[2 + (6 - 3)] - 4\} =$

Solve It

4. $\dfrac{\sqrt{19 - 3(6 - 8)}}{6[8 - 4(5 - 2)] - 1} =$

Solve It

---

5. $4 - 5|6 - 3(8 - 2)| =$

Solve It

6. $\dfrac{(9 - 1)5 - 4(6)}{\sqrt{11 - 2 - \sqrt{4 - \sqrt{2 + 7}}}} =$

Solve It

Distributing the Wealth

The *distributive property* is used to perform an operation on all the terms within a grouping symbol. The following rules show distributing multiplication over addition and distributing multiplication over subtraction.

$$a(b + c) = a \times b + a \times c \text{ and } a(b - c) = a \times b - a \times c$$

Q. 3(6 − 4) =

A. 6. First, distribute the 3 over the 6 minus 4: 3(6 − 4) = 3 × 6 − 3 × 4 = 18 − 12 = 6. Just subtracting the 4 from the 6 and then multiplying would be simpler, but I want to show you two ways to get the correct answer: 3(2) = 6. You generally use the distributive property when you can't combine what's in the grouping symbols.

Q. 5(a − ⅕) =

A. **5a − 1**

5 × a − 5 × ⅕ = 5a − 1

7. 4(7 + y) =

Solve It

8. −3(x − 11) =

Solve It

9. ⅔(6 + 15y) =

Solve It

10. −8(½ − ¼ + ⅜) =

Solve It

20 Part I: Getting a Grip on Basic Concepts

11. $4a(x - 2) =$

Solve It

12. $5(x + \frac{3}{5} - 2) =$

Solve It

Associating Correctly

The *associative rule* in math says that in addition and multiplication problems, you can change the *association* or groupings of three or more numbers and not change the final result. The associative rule looks like the following:

$$a + (b + c) = (a + b) + c \text{ and } a \times (b \times c) = (a \times b) \times c$$

This rule is special to addition and multiplication. It doesn't work for subtraction or division. You're probably wondering why even use this rule? Because it can sometimes make the computation easier.

- Instead of doing $5 + (-5 + 17)$, change it to $[5 + (-5)] + 17 = 0 + 17 = 17$.
- And instead of $6(\frac{1}{6} \times 19)$ do $(6 \times \frac{1}{6})19 = (1)19 = 19$.

Q. $4 + (5 + 6) =$

A. **15**. Add the terms in the parenthesis, first: $4 + (5 + 6) = 4 + 11 = 15$. Or you could re-associate the terms and add the first two together: $(4 + 5) + 6 = 9 + 6 = 15$.

Q. $4(5 \times 6) =$

A. **120**. Multiplying the way it's given, $4(5 \times 6) = 4(30) = 120$. Or you could re-associate and multiply the first two factors first: $(4 \times 5) 6 = (20)6 = 120$.

13. $16 + (-16 + 47) =$

Solve It

14. $(5 - 13) + 13 =$

Solve It

15. $18(\frac{5}{9} \times 7) =$

Solve It

16. $(110 \times 6)\frac{1}{6} =$

Solve It

Commuting to Work

The *commutative property* of addition and multiplication says that the order that you add or multiply numbers doesn't matter. However, the order of subtraction and division does matter. You get the same answer whether you multiple 3×4 or 4×3. The rule looks like the following:

$$a + b = b + a \textbf{ and } a \times b = b \times a$$

You can use this rule to your advantage when doing math computations. In the following two examples, the associative rule finishes off the problems after changing the order.

EXAMPLE

Q. $5 \times \frac{4}{7} \times \frac{1}{5} =$

A. $\frac{4}{7}$. You don't really want to multiply fractions unless necessary. Notice that the first and last factors are multiplicative inverses of one another: $5 \times \frac{4}{7} \times \frac{1}{5} = 5 \times \frac{1}{5} \times \frac{4}{7} = (5 \times \frac{1}{5})\frac{4}{7} = 1 \times \frac{4}{7} = \frac{4}{7}$. The second and last factors were reversed.

Q. $-3 + 16 + 303 =$

A. **316**. The first and second factors were reversed.

$$\begin{aligned} -3 + 16 + 303 &= 16 + (-3) + 303 \\ &= 16 + [-3 + 303] \\ &= 16 + 300 = 316 \end{aligned}$$

17. $8 + 5 + (-8) =$

Solve It

18. $5 \times 47 \times 2 =$

Solve It

19. ⅗ × 13 × 10 =

Solve It

20. −23 + 47 + 23 − 47 + 8 =

Solve It

21. ½ × 15 × 4 × ⅔ =

Solve It

22. −6 + 97 + 66 − 57 =

Solve It

Answers to Problems on Algebraic Properties

This section provides the answers (in bold) to the practice problems in this chapter.

1 $3(2-5)+14 = \mathbf{5}$

$3(2-5)+14 = 3(-3)+14 = (-9)+14 = 5$

2 $4[3(6-8)+2(5+9)]-11 = \mathbf{77}$

$$4[3(6-8)+2(5+9)]-11 = 4[3(-2)+2(14)]-11$$
$$= 4[-6+28]-11$$
$$= 4[22]-11 = 88-11 = 77$$

3 $5\{8[2+(6-3)]-4\} = \mathbf{180}$

$$5\{8[2+(6-3)]-4\} = 5\{8[2+3]-4\}$$
$$= 5\{8[5]-4\} = 5\{40-4\} = 5\{36\} = 180$$

4 $\dfrac{\sqrt{19-3(6-8)}}{6[8-4(5-2)]-1} = -\dfrac{1}{5}$

$$\dfrac{\sqrt{19-3(6-8)}}{6[8-4(5-2)]-1} = \dfrac{\sqrt{19-3(-2)}}{6[8-4(3)]-1} = \dfrac{\sqrt{19+6}}{6[8-12]-1} = \dfrac{\sqrt{25}}{6[-4]-1} = \dfrac{5}{-24-1} = \dfrac{5}{-25} = -\dfrac{1}{5}$$

5 $4-5|6-3(8-2)| = \mathbf{-56}$

$$4-5|6-3(8-2)| = 4-5|6-3(6)| = 4-5|6-18|$$
$$= 4-5|-12| = 4-5(12)$$
$$= 4-60 = -56$$

6 $\dfrac{(9-1)5-4(6)}{\sqrt{11-2}-\sqrt{4-\sqrt{2+7}}} = \mathbf{8}$

$$\dfrac{(9-1)5-4(6)}{\sqrt{11-2}-\sqrt{4-\sqrt{2+7}}} = \dfrac{(8)5-24}{\sqrt{9}-\sqrt{4-\sqrt{9}}} = \dfrac{40-24}{3-\sqrt{4-3}} = \dfrac{16}{3-\sqrt{1}} = \dfrac{16}{3-1} = \dfrac{16}{2} = 8$$

7 $4(7+y) = \mathbf{28+4y}$

$4(7+y) = 4\times 7 + 4\times y = 28+4y$

8 $-3(x-11) = \mathbf{-3x+33}$

$-3(x-11) = (-3)x-(-3)(11) = -3x+33$

9 $\frac{2}{3}(6+15y) = \mathbf{4+10y}$

$\frac{2}{3}(6+15y) = \frac{2}{3}\times 6 + \frac{2}{3}(15y) = \dfrac{2\times 6}{3} + \dfrac{2\times 15}{3}y = 4+10y$

10 $-8(\frac{1}{2}-\frac{1}{4}+\frac{3}{8}) = \mathbf{-5}$

$$-8\left(\dfrac{1}{2}-\dfrac{1}{4}+\dfrac{3}{8}\right) = (-8)\left(\dfrac{1}{2}\right)-(-8)\left(\dfrac{1}{4}\right)+(-8)\left(\dfrac{3}{8}\right) = -\dfrac{8}{2}+\dfrac{8}{4}-\dfrac{8\times 3}{8}$$
$$= -4+2-3$$
$$= (-4+2)-3 = -2-3 = -5$$

11 $4a(x-2) = \mathbf{4ax - 8a}$

$4a(x-2) = (4a) \times x - (4a)(2) = 4ax - 8a$

12 $5(x + \frac{4}{5} - 2) = \mathbf{5x - 6}$

$5\left(x + \frac{4}{5} - 2\right) = 5 \times x + 5\left(\frac{4}{5}\right) - 5(2) = 5x + \frac{5 \times 4}{5} - 10$
$= 5x + 4 - 10 = 5x - 6$

13 $16 + (-16 + 47) = \mathbf{47}$

$16 + (-16 + 47) = [16 + (-16)] + 47 = 0 + 47 = 47$

14 $(5 - 13) + 13 = \mathbf{5}$

$(5 - 13) + 13 = (5 + (-13)) + 13 = 5 + [(-13) + 13] = 5 + 0 = 5$

15 $18(\frac{5}{9} \times 7) = \mathbf{70}$

$18\left(\frac{5}{9} \times 7\right) = \left(18 \times \frac{5}{9}\right)7 = \left(\frac{18 \times 5}{9}\right)7 = (10)7 = 70$

16 $(110 \times 6)\frac{1}{6} = \mathbf{110}$

$(110 \times 6)\frac{1}{6} = 110(6 \times \frac{1}{6}) = 110(1) = 110$

17 $8 + 5 + (-8) = \mathbf{5}$

$8 + 5 + (-8) = 5 + 8 + (-8) = 5 + (8 - 8) = 5 + 0 = 5$

18 $5 \times 47 \times 2 = \mathbf{470}$

$5 \times 47 \times 2 = 5 \times 2 \times 47 = (5 \times 2)47 = 10 \times 47 = 470$

19 $\frac{3}{5} \times 13 \times 10 = \mathbf{78}$

$\frac{3}{5} \times 13 \times 10 = \frac{3}{5} \times 10 \times 13 = \left(\frac{3}{5} \times 10\right)13 = \left(\frac{3 \times 10}{5}\right)13 = 6 \times 13 = 78$

20 $-23 + 47 + 23 - 47 + 8 = \mathbf{8}$

$-23 + 47 + 23 - 47 + 8 = -23 + 23 + 47 - 47 + 8$
$= (-23 + 23) + (47 - 47) + 8$
$= 0 + 0 + 8 = 8$

21 $\frac{1}{2} \times 15 \times 4 \times \frac{2}{3} = \mathbf{20}$

$\frac{1}{2} \times 15 \times 4 \times \frac{2}{3} = \frac{1}{2} \times 4 \times 15 \times \frac{2}{3} = \left(\frac{1}{2} \times 4\right)\left(15 \times \frac{2}{3}\right) = \left(\frac{4}{2}\right)\left(\frac{15 \times 2}{3}\right) = 2 \times 10 = 20$

22 $-6 + 97 + 66 - 57 = \mathbf{100}$

$-6 + 97 + 66 - 57 = -6 + 66 + 97 - 57 = (-6 + 66) + (97 - 57)$
$= 60 + 40 = 100$

Chapter 3
Working with Fractions

In This Chapter
- Simplifying and changing fractions
- Making use of proportions
- Operating on fractions

You can try to run and hide, but you may as well face it. Fractions are here to stay. People don't usually eat a whole pizza or buy furniture that's exactly five feet long or wear a hat that's exactly a whole number of inches in size. Fractions are an essential part of everyday life.

Fractions are just as important in algebra. In order to complete a problem, you have to know how to switch from one form to another. This chapter provides you plenty of questions to work out all your fractions frustration.

Converting Improper and Mixed Fractions

An *improper fraction* is one where the *numerator* (the number on the top of the fraction) has a greater value than the *denominator* (the number on the bottom of the fraction) — the fraction is top heavy. Improper fractions can be written as *mixed numbers* — the whole number part that tells how many times the denominator divides the numerator, and the fraction part that gives the remainder in the division. For example, $4\frac{2}{3}$ is a mixed number.

To change an improper fraction to a mixed number, divide the numerator by the denominator, and write the remainder in the numerator of the new fraction.

To change a mixed number to an improper fraction, multiply the whole number times the denominator and add the numerator. This result goes in the numerator of a fraction that has the original denominator still in the denominator.

Q. Change $\frac{29}{8}$ to a mixed number.

A. $3\frac{5}{8}$. First, divide the 29 by 8:

$$8\overline{)\begin{array}{r}3\\29\\\underline{24}\\5\end{array}}$$

Then write the mixed number with the quotient as the whole number and the remainder as the numerator of the fraction in the mixed number: $3\frac{5}{8}$

Q. Change $6\frac{5}{7}$ to an improper fraction.

A. $\frac{47}{7}$. Multiply the 6 and 7 and then add the 5, which equals 47. Then write the fraction with this result in the numerator and the 7 in the denominator: $\frac{47}{7}$

Part I: Getting a Grip on Basic Concepts

1. Change the mixed number to an improper fraction: $4\frac{7}{8}$

    Solve It

2. Change the mixed number to an improper fraction: $2\frac{1}{3}$

    Solve It

3. Change the mixed number to an improper fraction: $-5\frac{3}{5}$

    Solve It

4. Change the improper fraction to a mixed number: $\frac{9}{5}$

    Solve It

5. Change the improper fraction to a mixed number: $\frac{19}{7}$

    Solve It

6. Change the improper fraction to a mixed number: $\frac{280}{11}$

    Solve It

Finding Equivalent Fractions

In algebra, fractions are used in all sorts of computations and manipulations. In many of these processes, you have to change the fractions so that they have the same denominator or so their form is compatible with what you need to solve the problem. Two fractions are *equivalent* if they have the same value, such as ½ and ⅜. To create an equivalent fraction from a given fraction, you multiply or divide the numerator and denominator by the same number.

To reduce a fraction, multiply the numerator and denominator by the same number.

Q. Find a fraction with a denominator of 40 for ⅞.

A. $\frac{35}{40}$. You multiply 5 times 8 to get 40; you also have to multiply the numerator by the 5. Basically, you're just multiplying by 1; multiplying by 1 doesn't change the real value of anything.

$$\frac{7}{8} \times \frac{5}{5} = \frac{35}{40}$$

Q. Reduce $\frac{15}{36}$ by multiplying the numerator and denominator by ⅓.

A. $\frac{5}{12}$.

$$\frac{15}{36} \times \frac{\frac{1}{3}}{\frac{1}{3}} = \frac{5}{12}$$

7. Find an equivalent fraction with a denominator of 28: ¾

Solve It

8. Find an equivalent fraction with a denominator of 24: ⅝

Solve It

9. Find an equivalent fraction with a denominator of 30: ⅚

Solve It

10. Find an equivalent fraction with a denominator of 60: $\frac{2}{15}$

Solve It

28 Part I: Getting a Grip on Basic Concepts

11. Reduce the fraction: $\frac{16}{90}$.

Solve It

12. Reduce the fraction: $\frac{63}{84}$

Solve It

Making It Proportional

A *proportion* is an equation with two fractions equal to one another. A proportion has some wonderful properties that make it useful for solving problems when you're comparing one quantity to another or one percentage to another.

Given the proportion $\frac{a}{b} = \frac{c}{d}$, then the "flip" of it (or $\frac{b}{a} = \frac{d}{c}$) is also true. You can cross-multiply and get a true statement: $a \times d = c \times b$. You can reduce the fractions as usual, finding a common factor for the numerator and denominator of one of the fractions. You can also reduce going across, too. If a and c have a common factor, you can divide each by that factor. The same goes for b and d.

Q. Find the missing value in the following proportion:

$$\frac{42}{66} = \frac{28}{d}$$

A. 44. The numerator and denominator in the fraction on the left have a common factor of 6. Multiply each by $\frac{1}{6}$. Flip the proportion to get the unknown in the numerator. Then you see that the two bottom numbers each have a common factor of 7. Multiply each by $\frac{1}{7}$. Finally, cross-multiply to get your answer:

$$\frac{42}{66} \times \frac{\frac{1}{6}}{\frac{1}{6}} = \frac{7}{11}$$

$$\frac{7}{11} = \frac{28}{d}$$

$$\frac{11}{7} = \frac{d}{28}$$

$$\frac{11}{7_1} = \frac{d}{28_4}$$

$$1 \times d = 4 \times 11$$

$$d = 44$$

Q. What's the state capital of Vermont?

A. Montpelier. What? Geography? Well, I didn't want you to get so enthralled with your algebra problems that you forget where your next vacation might be. And, where else can you drink some fresh apple cider, try some maple syrup, and visit the nearby famous ice cream factory started by two environmentally friendly guys with crazy flavors? (I don't know about you, but suddenly I'm starving.)

13. Solve for x: $\frac{7}{21} = \frac{x}{24}$

Solve It

14. Solve for x: $\frac{45}{x} = \frac{60}{200}$

Solve It

15. Solve for x: $\frac{x}{90} = \frac{60}{108}$

Solve It

16. Solve for x: $\frac{26}{16} = \frac{65}{x}$

Solve It

17. A recipe calls for 2 teaspoons of cinnamon and 4 cups of flour. You need to increase the flour to 6 cups. To keep this proportional, how many teaspoons of cinnamon should you use? *Hint:* Fill in the proportion $\frac{\text{original cinnamon}}{\text{original flour}} = \frac{\text{new cinnamon}}{\text{new flour}}$ and let x represent the new cinnamon.

Solve It

18. A factory produces two faulty televisions for every 500 televisions it produces. How many faulty televisions would you expect to find in a shipment of 1,250 televisions?

Solve It

Finding Common Denominators

Before you can add or subtract fractions, you need to find a common denominator. Ideally that common denominator is the *least common multiple* — the smallest number that each of the denominators can divide into. A method of last resort, though, is to multiply the denominators together. Doing so gives you a number they divide evenly. Then you can refine your answer by dividing by the greatest common factor — the largest number that divides the numbers evenly.

Q. How would you write the fractions $\frac{1}{3}$ and $\frac{3}{4}$ with the same denominator?

A. $\frac{4}{12}$ **and** $\frac{9}{12}$. The fractions $\frac{1}{3}$ and $\frac{3}{4}$ have denominators with nothing in common, so the common denominator is 12, the product of the two numbers. Now you can write them both as fractions with a denominator of 12: $\frac{1}{3} \times \frac{4}{4} = \frac{4}{12}$ and $\frac{3}{4} \times \frac{3}{3} = \frac{9}{12}$

Q. What is the least common denominator for the fractions $\frac{7}{20}$ and $\frac{5}{24}$?

A. **120.** The fractions $\frac{7}{20}$ and $\frac{5}{24}$ have denominators that have a greatest common factor of 4. So $\frac{7}{20} \times \frac{6}{6} = \frac{42}{120}$ and $\frac{5}{24} \times \frac{5}{5} = \frac{25}{120}$. You can multiply the two denominators together and then divide by that common factor: $20 \times 24 = 480$ and $480 \div 4 = 120$. The least common denominator is 120.

19. Rewrite the fractions with a common denominator: ⅔ and ⅜

Solve It

20. Rewrite the fractions with a common denominator: 5/12 and 7/18

Solve It

21. Rewrite the fractions with a common denominator: 3/10 and 49/60

Solve It

22. Rewrite the fractions with a common denominator: ¾ and ⅙

Solve It

32 Part I: Getting a Grip on Basic Concepts

23. Rewrite the fractions with a common denominator: ½, ⅓, and ⅙

Solve It

24. Rewrite the fractions with a common denominator: ⅔, ⅚, and ¾

Solve It

Adding Fractions

You can add fractions together if they have a common denominator. After you find the common denominator and change the fractions to their equivalents, then you can add them.

Q. $\frac{5}{6} + \frac{7}{8} =$

A. $1\frac{17}{24}$. First find the common denominator of 24, and then complete the addition:

$$\frac{5}{6} \times \frac{4}{4} + \frac{7}{8} \times \frac{3}{3} = \frac{20}{24} + \frac{21}{24} = \frac{41}{24} = 1\frac{17}{24}$$

Q. $2½ + (-1⅓) + 5\frac{3}{10} =$

A. $6\frac{7}{15}$. You need the common denominator of 30:

$$\frac{5}{2} + \left(-\frac{4}{3}\right) + \frac{53}{10} = \frac{5}{2} \times \frac{15}{15} + \left(-\frac{4}{3}\right) \times \frac{10}{10} + \frac{53}{10} \times \frac{3}{3}$$

$$= \frac{75}{30} + \left(-\frac{40}{30}\right) + \frac{159}{30}$$

$$= \frac{194}{30}$$

$$= 6\frac{14}{30}$$

$$= 6\frac{7}{15}$$

25. $\frac{2}{5} + \frac{1}{2} =$

Solve It

26. $\frac{3}{8} + \frac{7}{12} =$

Solve It

27. $\frac{9}{10} + \frac{2}{45} =$

Solve It

28. $3\frac{1}{3} + 4\frac{3}{5} + \frac{7}{15} =$

Solve It

Part I: Getting a Grip on Basic Concepts

Subtracting Fractions

Subtracting fractions involves the same requirement as adding fractions — the denominators have to be the same.

Q. $2\frac{1}{8} - 1\frac{1}{7} =$

A. $\frac{55}{56}$. Find the common denominator of 56:

$$\frac{17}{8} - \frac{8}{7} = \frac{17}{8} \times \frac{7}{7} - \frac{8}{7} \times \frac{8}{8} = \frac{119}{56} - \frac{64}{56} = \frac{55}{56}$$

Q. $1 - \frac{11}{13} =$

A. $\frac{2}{13}$. Even though the 1 isn't a fraction, you need to write it as a fraction with a denominator of 13. The subtraction problem becomes $\frac{13}{13} - \frac{11}{13}$.

29. $\frac{2}{3} - \frac{1}{6} =$

Solve It

30. $1\frac{5}{12} - \frac{7}{8} =$

Solve It

31. $3\frac{1}{3} - 2\frac{1}{5} =$

Solve It

32. $3\frac{2}{3} - (-6\frac{1}{2}) =$

Solve It

Multiplying Fractions

Multiplying fractions is really a much easier process than adding or subtracting fractions because you don't have to find a common denominator. Furthermore, you can take some creative steps and reduce the fractions before you ever multiply them.

You can divide the numerator of any fraction and the denominator of any other fraction by the same number. Doing so saves your having large numbers to multiply and then to reduce later.

When you start with mixed numbers, you have to change them to improper fractions before starting the reduction and multiplication process.

Q. $\dfrac{25}{14} \times \dfrac{49}{30} \times \dfrac{27}{10}$

A. $7\tfrac{7}{8}$. First, reduce: the 25 and 30 have a common factor of 5, the 14 and 49 have a common factor of 7, the 6 and 27 have a common factor of 3, and the 5 and 10 have a common factor of 5. And then, to multiply the fractions, multiply all the numerators together and all the denominators together to make the new fraction:

$$\dfrac{\overset{5}{\cancel{25}}}{14} \times \dfrac{49}{\underset{6}{\cancel{30}}} \times \dfrac{27}{10} = \dfrac{\overset{5}{\cancel{25}}}{\underset{2}{\cancel{14}}} \times \dfrac{\overset{7}{\cancel{49}}}{\underset{6}{\cancel{30}}} \times \dfrac{27}{10}$$

$$= \dfrac{\overset{5}{\cancel{25}}}{\underset{2}{\cancel{14}}} \times \dfrac{\overset{7}{\cancel{49}}}{\underset{\cancel{6}_2}{\cancel{30}}} \times \dfrac{\overset{9}{\cancel{27}}}{10}$$

$$= \dfrac{\overset{\cancel{5}^1}{\cancel{25}}}{\underset{2}{\cancel{14}}} \times \dfrac{\overset{7}{\cancel{49}}}{\underset{\cancel{6}_2}{\cancel{30}}} \times \dfrac{\overset{9}{\cancel{27}}}{\underset{2}{\cancel{10}}}$$

$$= \dfrac{1}{2} \times \dfrac{7}{2} \times \dfrac{9}{2}$$

$$= \dfrac{63}{8}$$

$$= 7\dfrac{7}{8}$$

Q. $-2\tfrac{2}{9} \times 1\tfrac{1}{8} =$

A. $-2\tfrac{1}{2}$. $-2\dfrac{2}{9} \times 1\dfrac{1}{8} = -\dfrac{20}{9} \times \dfrac{9}{8}$

$$= -\dfrac{20}{\underset{1}{\cancel{9}}} \times \dfrac{\overset{1}{\cancel{9}}}{8}$$

$$= -\dfrac{\overset{5}{\cancel{20}}}{\underset{1}{\cancel{9}}} \times \dfrac{\overset{1}{\cancel{9}}}{\underset{2}{\cancel{8}}}$$

$$= -\dfrac{5}{2}$$

$$= -2\dfrac{1}{2}$$

---

33. $\dfrac{10}{21} \times \dfrac{14}{15} =$

Solve It

34. $\dfrac{6}{11} \times \left(-\dfrac{10}{21}\right) \times \dfrac{77}{25} =$

Solve It

Part I: Getting a Grip on Basic Concepts

35. $4\frac{1}{5} \times \frac{25}{49} =$

Solve It

36. $\frac{x}{15} \times \frac{10}{11} =$

Solve It

37. $2\frac{2}{3} \times 1\frac{4}{5} \times \frac{7}{24} =$

Solve It

38. $\left(-\frac{7}{27}\right) \times \frac{18}{25} \times \left(-\frac{15}{28}\right) =$

Solve It

Chapter 3: Working with Fractions   **37**

Dividing Fractions

Algebra really doesn't have a way to divide fractions. If you want to divide fractions, you first have to change them to multiplication problems. Sounds easy, right? Just change the division to multiplication and use the *reciprocal* — where the numerator and denominator switch places — of the second fraction in that new problem. The answer to this multiplication problem is the same as the answer to the original division problem.

Q. $6\frac{1}{8} \div 5\frac{1}{4} =$

A. $1\frac{1}{6}$. First change the mixed numbers to improper fractions. Then change the divide to multiply and change the second (right) fraction to its reciprocal. Then do the multiplication problem to get the answer:

$$\frac{49}{8} \div \frac{21}{4} = \frac{49}{8} \times \frac{4}{21} = \frac{\cancel{49}^{7}}{\cancel{8}_{2}} \times \frac{\cancel{4}^{1}}{\cancel{21}_{3}} = \frac{7}{6} = 1\frac{1}{6}$$

Q. $2 \div \frac{3}{5} =$

A. $3\frac{1}{3}$. First change the 2 to a fraction: $\frac{2}{1}$. Then change the divide to multiply and change the second (right) fraction to its reciprocal. Then do the multiplication problem to get the answer: $\frac{10}{3}$ which can be changed to the mixed number.

$$\frac{2}{1} \times \frac{5}{3} = \frac{10}{3} = 3\frac{1}{3}$$

39. $\frac{4}{7} \div \frac{2}{3} =$

Solve It

40. $-\frac{15}{14} \div \left(-\frac{20}{21}\right) =$

Solve It

41. $2\frac{1}{2} \div \frac{3}{4} =$

Solve It

42. $-8\frac{1}{3} \div 1\frac{1}{3} =$

Solve It

43. $\frac{121}{36} \div \left(-\frac{22}{45}\right) =$

Solve It

44. $7\frac{1}{2} \div 3\frac{3}{4} =$

Solve It

Changing Fractions to Decimals and Vice Versa

Every fraction with an integer in the numerator and denominator has a decimal equivalent. Sometimes these decimals end (terminate); sometimes the decimals go on forever, repeating a pattern of numbers over and over.

To change a fraction to a decimal, divide the denominator into the numerator, adding zeros after the decimal point until the division problem either ends, or it has a pattern. To indicate a pattern repeating over and over, draw a line across the top of the digits that repeat (for example, $.2\overline{333}$).

To change a decimal to a fraction, place the digits of a terminating decimal over a power of 10 with as many zeros as there are decimal values. Then reduce the fraction. To change a repeating decimal to a fraction (this tip works only for those repeating decimals where all the digits repeat, from the beginning), place the repeating digits in the numerator of a fraction that has nines in the denominator. It should have as many nines as digits that repeat. For instance, in the repeating decimal .123123123 . . . you'd put 123 over 999 and then reduce the fraction.

Q. Change 0.452 to a fraction.

A. $^{113}/_{250}$. Put 452 over 1,000 and reduce.

Q. Change 0.00016 to a fraction.

A. $^{1}/_{6250}$. Put 16 over 100,000 and reduce.

Q. Change $0.\overline{18}$ to a fraction.

A. $^{2}/_{11}$. Put 18 over 99 and reduce.

Q. Change $0.\overline{285714}$ to a fraction.

A. $^{2}/_{7}$. Put 285714 over 999999 and reduce.

---

45. Change ⅜ to a decimal.

Solve It

46. Change ⁴⁄₉ to a decimal.

Solve It

47. Change 3/11 to a decimal.

Solve It

48. Change 0.45 to a fraction.

Solve It

49. Change $0.3\overline{6}$ to a fraction.

Solve It

50. Change $0.\overline{405}$ to a fraction.

Solve It

Answers to Problems on Fractions

This section provides the answers (in bold) to the practice problems in this chapter.

1 Change the mixed number to an improper fraction: $4\frac{7}{8}$. The answer is $\frac{39}{8}$. $4 \times 8 + 7 = 32 + 7 = 39$ so the improper fraction is $\frac{39}{8}$.

2 Change the mixed number to an improper fraction: $2\frac{1}{13}$. The answer is $\frac{27}{13}$. $2 \times 13 + 1 = 26 + 1 = 27$ so the improper fraction is $\frac{27}{13}$.

3 Change the mixed number to an improper fraction: $-5\frac{3}{5}$. The answer is $-\frac{28}{5}$.

$$-5\frac{3}{5} = -\left(5\frac{3}{5}\right)$$
$$5 \times 5 + 3 = 25 + 3 = 28$$

So $5\frac{3}{5} = \frac{28}{5}$ and $-5\frac{3}{5} = -\frac{28}{5}$.

4 Change the improper fraction to a mixed number: $\frac{6}{5}$. The answer is $1\frac{1}{5}$. $5\overline{)6}$ with quotient 1, $\frac{5}{1}$.

Think of breaking up the fraction into two pieces that are equal to $\frac{6}{5}$, which equals $1\frac{1}{5}$.

5 Change the improper fraction to a mixed number: $\frac{19}{7}$. The answer is $2\frac{5}{7}$. $7\overline{)19}$ quotient 2, $\frac{14}{5}$.

Think of breaking up the fraction into two pieces that are equal to $\frac{19}{7}$, which equals $2\frac{5}{7}$.

6 Change the improper fraction to a mixed number: $\frac{280}{11}$. The answer is $25\frac{5}{11}$. $11\overline{)280}$ quotient 25, $\frac{22}{60}$, $\frac{55}{5}$.

Think of breaking up the fraction into two pieces that result in $\frac{280}{11}$, which equals $25\frac{5}{11}$.

7 Find an equivalent fraction with a denominator of 28: $\frac{3}{7}$. The answer is $\frac{12}{28}$. To get 28 in the denominator, multiply 7 by 4. $\frac{3}{7} = \frac{3}{7} \times \frac{4}{4} = \frac{12}{28}$

8 Find an equivalent fraction with a denominator of 24: $\frac{5}{8}$. The answer is $\frac{15}{24}$. $\frac{5}{8} = \frac{5}{8} \times \frac{3}{3} = \frac{15}{24}$

9 Find an equivalent fraction with a denominator of 30: $\frac{x}{6}$. The answer is $\frac{5x}{30}$. $\frac{x}{6} = \frac{x}{6} \times \frac{5}{5} = \frac{5x}{30}$

10 Find an equivalent fraction with a denominator of 60: $\frac{2x}{15}$. The answer is $\frac{8x}{60}$. $\frac{2x}{15} = \frac{2x}{15} \times \frac{4}{4} = \frac{8x}{60}$

11 Reduce the fraction: $\frac{16}{60}$. The answer is $\frac{4}{15}$. 4 is the largest common divisor of 16 and 60, because $16 = 4 \times 4$ and $60 = 15 \times 4$. So multiply $\frac{16}{60}$ by $\frac{\frac{1}{4}}{\frac{1}{4}}$.

You get $\frac{16}{60} = \frac{16}{60} \times \frac{\frac{1}{4}}{\frac{1}{4}} = \frac{16 \times \frac{1}{4}}{60 \times \frac{1}{4}} = \frac{4}{15}$.

12 Reduce the fraction: $\frac{63}{84}$. The answer is ¾. 21 is the largest common divisor of 63 and 84, because 63 = 3 × 21 and 84 = 4 × 21. So $\frac{63}{84} = \frac{63}{84} \times \frac{\frac{1}{21}}{\frac{1}{21}} = \frac{63 \times \frac{1}{21}}{84 \times \frac{1}{21}} = \frac{3}{4}$.

13 Solve for x: $\frac{7}{21} = \frac{x}{24}$. The answer is $x = 8$.

$$\frac{7}{21} = \frac{x}{24} \rightarrow \frac{7}{21} \times \frac{\frac{1}{7}}{\frac{1}{7}} = \frac{x}{24} \rightarrow \frac{1}{3} = \frac{x}{24} \rightarrow x = \frac{1}{3} \times 24 \rightarrow x = 8$$

14 Solve for x: $\frac{45}{x} = \frac{60}{200}$. The answer is $x = 150$.

$$\frac{45}{x} = \frac{60}{200} \rightarrow \frac{45}{x} = \frac{60}{200} \times \frac{\frac{1}{20}}{\frac{1}{20}} \rightarrow \frac{45}{x} = \frac{3}{10} \rightarrow \frac{x}{45} = \frac{10}{3} \rightarrow x = \frac{10}{3} \times 45 = 150$$

15 Solve for x: $\frac{x}{90} = \frac{60}{108}$. The answer is $x = 50$.

$$\frac{x}{90} = \frac{60}{108} \rightarrow \frac{x}{90} = \frac{60}{108} \times \frac{\frac{1}{12}}{\frac{1}{12}} \rightarrow \frac{x}{90} = \frac{5}{9} \rightarrow x = \frac{5}{9} \times 90 = 50$$

16 Solve for x: $\frac{26}{16} = \frac{65}{x}$. The answer is $x = 40$.

$$\frac{26}{16} = \frac{65}{x} \rightarrow \frac{26}{16} \times \frac{\frac{1}{2}}{\frac{1}{2}} = \frac{65}{x} \rightarrow \frac{13}{8} = \frac{65}{x} \rightarrow \frac{8}{13} = \frac{x}{65} \rightarrow x = \frac{8}{13} \times 65 = 40$$

17 A recipe calls for 2 teaspoons of cinnamon and 4 cups of flour. You need to increase the flour to 6 cups. To keep this proportional, how many teaspoons of cinnamon should you use? *Hint:* Fill in the proportion $\frac{\text{original cinnamon}}{\text{original flour}} = \frac{\text{new cinnamon}}{\text{new flour}}$ and let x represent the new cinnamon. The answer is $x = 3$.

$$\frac{2}{4} = \frac{x}{6} \rightarrow \frac{2}{4} \times \frac{\frac{1}{2}}{\frac{1}{2}} = \frac{x}{6} \rightarrow \frac{1}{2} = \frac{x}{6} \rightarrow x = \frac{1}{2} \times 6 = 3$$

18 A factory produces two faulty televisions for every 500 televisions it produces. How many faulty televisions would you expect to find in a shipment of 1,250 televisions? The answer is $x = 5$.

$$\frac{2 \text{ faulty tv's}}{500 \text{ tv's}} = \frac{x \text{ faulty tv's}}{1250 \text{ tv's}} \rightarrow \frac{2}{500} = \frac{x}{1250}$$

$$\rightarrow \frac{2}{500} \times \frac{\frac{1}{2}}{\frac{1}{2}} = \frac{x}{1250}$$

$$\rightarrow \frac{1}{250} = \frac{x}{1250} \rightarrow x = 1250 \times \frac{1}{250} = 5$$

19 Rewrite the fractions with a common denominator: $\frac{2}{7}$ and $\frac{3}{8}$. The answers are $\frac{16}{56}$ **and** $\frac{21}{56}$. The largest common factor of 7 and 8 is 1. So the least common denominator is 56 ($\frac{7 \times 8}{1} = 56$). Here are the details: $\frac{2}{7} = \frac{2}{7} \times \frac{8}{8} = \frac{16}{56}$ and $\frac{3}{8} = \frac{3}{8} \times \frac{7}{7} = \frac{21}{56}$.

Chapter 3: Working with Fractions   **43**

20 Rewrite the fractions with a common denominator: $5/12$ and $7/18$. The answers are **$15/36$ and $14/36$**. The largest common factor of 12 and 18 is 6. The least common denominator is 36 ($\frac{12 \times 18}{6} = 36$). Here are the details: $\frac{5}{12} = \frac{5}{12} \times \frac{3}{3} = \frac{15}{36}$ and $\frac{7}{18} = \frac{7}{18} \times \frac{2}{2} = \frac{14}{36}$

21 Rewrite the fractions with a common denominator: $9/10$ and $49/60$. The answers are **$54/60$ and $49/60$**. The largest common factor of 10 and 60 is 10. The least common denominator is 60 ($\frac{10 \times 60}{10} = 60$). Break it down: $\frac{9}{10} = \frac{9}{10} \times \frac{6}{6} = \frac{54}{60}$ and $49/60$ remains as is.

22 Rewrite the fractions with a common denominator: $5/x$ and $1/6$. The answers are **$30/6x$ and $x/6x$**. The largest common factor of x and 6 is 1. The least common denominator of them is $6x$ ($\frac{6 \times x}{1} = 6x$). Here's the long of it: $\frac{5}{x} = \frac{5}{x} \times \frac{6}{6} = \frac{30}{6x}$ and $\frac{1}{6} = \frac{1}{6} \times \frac{x}{x} = \frac{x}{6x}$

23 Rewrite the fractions with a common denominator: $1/2$ and $1/3$ and $1/5$. The answers are **$15/30$, $10/30$, and $6/30$**. The least common denominator of fractions with denominators of 2, 3, and 5 is 30. Write it out:
$\frac{1}{2} = \frac{1}{2} \times \frac{15}{15} = \frac{15}{30}$, $\frac{1}{3} = \frac{1}{3} \times \frac{10}{10} = \frac{10}{30}$, and $\frac{1}{5} = \frac{1}{5} \times \frac{6}{6} = \frac{6}{30}$

24 Rewrite the fractions with a common denominator: $2/3$, $5/6$, and $3/4$. The answers are **$8/12$, $10/12$, and $9/12$**. The least common denominator of $2/3$ and $5/6$ is 6. And the least common denominator of a fraction with a denominator of 6 and 4 is 12 ($\frac{6 \times 4}{2} = 12$). In long hand:
$\frac{2}{3} = \frac{2}{3} \times \frac{4}{4} = \frac{8}{12}$, $\frac{5}{6} = \frac{5}{6} \times \frac{2}{2} = \frac{10}{12}$, and $\frac{3}{4} = \frac{3}{4} \times \frac{3}{3} = \frac{9}{12}$

25 $\frac{2}{5} + \frac{1}{2} =$ The least common denominator is 10, so the answer is **$9/10$**.
$\frac{2}{5} + \frac{1}{2} = \frac{2}{5} \times \frac{2}{2} + \frac{1}{2} \times \frac{5}{5} = \frac{4}{10} + \frac{5}{10} = \frac{9}{10}$

26 $\frac{3}{8} + \frac{7}{12} =$ The least common denominator is 24, so the answer is **$23/24$**.
$\frac{3}{8} + \frac{7}{12} = \frac{3}{8} \times \frac{3}{3} + \frac{7}{12} \times \frac{2}{2} = \frac{9}{24} + \frac{14}{24} = \frac{23}{24}$

27 $\frac{9}{10} + \frac{2}{45} =$ The least common denominator is 90. ($\frac{10 \times 45}{5} = 90$), so the answer is **$17/18$**.
$\frac{9}{10} + \frac{2}{45} = \frac{9}{10} \times \frac{9}{9} + \frac{2}{45} \times \frac{2}{2} = \frac{81}{90} + \frac{4}{90} = \frac{85}{90} = \frac{85}{90} \times \frac{\frac{1}{5}}{\frac{1}{5}} = \frac{17}{18}$

28 $3\frac{1}{3} + 4\frac{3}{5} + 7/15 =$ The least common denominator is 15, so the answer is **$8\frac{2}{5}$**.
$3\frac{1}{3} + 4\frac{3}{5} + \frac{7}{15} = \frac{10}{3} + \frac{23}{5} + \frac{7}{15} = \frac{10}{3} \times \frac{5}{5} + \frac{23}{5} \times \frac{3}{3} + \frac{7}{15}$
$= \frac{50}{15} + \frac{69}{15} + \frac{7}{15} = \frac{126}{15}$
$= \frac{126}{15} \times \frac{\frac{1}{3}}{\frac{1}{3}} = \frac{42}{5} = 8\frac{2}{5}$

29 $\frac{2}{3} - \frac{1}{6} = \frac{1}{2}$
$\frac{2}{3} - \frac{1}{6} = \frac{2}{3} \times \frac{2}{2} - \frac{1}{6} = \frac{4}{6} - \frac{1}{6} = \frac{3}{6} = \frac{1}{2}$

30 $1\frac{5}{12} - \frac{7}{9} = \frac{23}{36}$

$1\frac{5}{12} - \frac{7}{9} = \frac{17}{12} - \frac{7}{9} = \frac{17}{12} \times \frac{3}{3} - \frac{7}{9} \times \frac{4}{4} = \frac{51}{36} - \frac{28}{36} = \frac{23}{36}$

31 $3\frac{1}{3} - 2\frac{4}{5} = \frac{8}{15}$

$3\frac{1}{3} - 2\frac{4}{5} = \frac{10}{3} - \frac{14}{5} = \frac{10}{3} \times \frac{5}{5} - \frac{14}{5} \times \frac{3}{3} = \frac{50}{15} - \frac{42}{15} = \frac{8}{15}$

32 $3\frac{2}{3} - (-6\frac{1}{2}) = 10\frac{1}{6}$

$3\frac{2}{3} - \left(-6\frac{1}{2}\right) = 3\frac{2}{3} + 6\frac{1}{2} = \frac{11}{3} + \frac{13}{2} = \frac{11}{3} \times \frac{2}{2} + \frac{13}{2} \times \frac{3}{3} = \frac{22}{6} + \frac{39}{6} = \frac{61}{6} = 10\frac{1}{6}$

33 $\frac{10}{21} \times \frac{14}{15} = \frac{4}{9}$

$\frac{10}{21} \times \frac{14}{15} = \frac{10 \times 14}{21 \times 15} = \frac{\overset{2}{\cancel{10}} \times \overset{2}{\cancel{14}}}{\underset{3}{\cancel{21}} \times \underset{3}{\cancel{15}}} = \frac{2 \times 2}{3 \times 3} = \frac{4}{9}$

34 $\frac{6}{11} \times \left(-\frac{10}{21}\right) \times \frac{77}{25} = -\frac{4}{5}$

$\frac{6}{11} \times \left(-\frac{10}{21}\right) \times \frac{77}{25} = -\frac{\overset{2}{\cancel{6}} \times \overset{2}{\cancel{10}} \times \overset{7}{\cancel{77}}}{\underset{1}{\cancel{11}} \times \underset{7}{\cancel{21}} \times \underset{5}{\cancel{25}}} = -\frac{2 \times 2 \times \overset{1}{\cancel{7}}}{1 \times \underset{1}{\cancel{7}} \times 5} = -\frac{2 \times 2}{1 \times 5} = -\frac{4}{5}$

35 $4\frac{1}{5} \times \frac{25}{49} = 2\frac{1}{7}$

$4\frac{1}{5} \times \frac{25}{49} = \frac{21}{5} \times \frac{25}{49} = \frac{\overset{3}{\cancel{21}} \times \overset{5}{\cancel{25}}}{\underset{1}{\cancel{5}} \times \underset{7}{\cancel{49}}} = \frac{3 \times 5}{1 \times 7} = \frac{15}{7} = 2\frac{1}{7}$

36 $\frac{x}{15} \times \frac{10}{11} = \frac{2x}{33}$

$\frac{x}{15} \times \frac{10}{11} = \frac{x \times \overset{2}{\cancel{10}}}{\underset{3}{\cancel{15}} \times 11} = \frac{x \times 2}{3 \times 11} = \frac{2x}{33}$

37 $2\frac{2}{3} \times 1\frac{4}{5} \times \frac{7}{24} = 1\frac{2}{5}$

$2\frac{2}{3} \times 1\frac{4}{5} \times \frac{7}{24} = \frac{8}{3} \times \frac{9}{5} \times \frac{7}{24} = \frac{\overset{1}{\cancel{8}} \times \overset{3}{\cancel{9}} \times 7}{\underset{1}{\cancel{3}} \times 5 \times \underset{3}{\cancel{24}}} = \frac{1 \times \overset{1}{\cancel{3}} \times 7}{1 \times 5 \times \underset{1}{\cancel{3}}} = \frac{7}{5} = 1\frac{2}{5}$

38 $\left(-\frac{7}{27}\right) \times \frac{18}{25} \times \left(-\frac{15}{28}\right) = \frac{1}{10}$

$\left(-\frac{7}{27}\right) \times \frac{18}{25} \times \left(-\frac{15}{28}\right) = +\frac{\overset{1}{\cancel{7}} \times \overset{2}{\cancel{18}} \times \overset{3}{\cancel{15}}}{\underset{3}{\cancel{27}} \times \underset{5}{\cancel{25}} \times \underset{4}{\cancel{28}}} = \frac{1 \times \overset{1}{\cancel{2}} \times \overset{1}{\cancel{3}}}{\underset{1}{\cancel{3}} \times 5 \times \underset{2}{\cancel{4}}} = \frac{1 \times 1 \times 1}{1 \times 5 \times 2} = \frac{1}{10}$

39 $\frac{4}{7} \div \frac{2}{3} = \frac{6}{7}$

$\frac{4}{7} \div \frac{2}{3} = \frac{4}{7} \times \frac{3}{2} = \frac{4 \times 3}{7 \times 2} = \frac{\overset{2}{\cancel{4}} \times 3}{7 \times \underset{1}{\cancel{2}}} = \frac{2 \times 3}{7 \times 1} = \frac{6}{7}$

40 $-\frac{15}{14} \div \left(-\frac{20}{21}\right) = 1\frac{1}{8}$

$-\frac{15}{14} \div \left(-\frac{20}{21}\right) = -\frac{15}{14} \times \left(-\frac{21}{20}\right) = \frac{15}{14} \times \frac{21}{20} = \frac{\overset{3}{\cancel{15}} \times \overset{3}{\cancel{21}}}{\underset{2}{\cancel{14}} \times \underset{4}{\cancel{20}}} = \frac{3 \times 3}{2 \times 4} = \frac{9}{8} = 1\frac{1}{8}$

41 $2\frac{1}{2} \div \frac{3}{4} = 3\frac{1}{3}$

$2\frac{1}{2} \div \frac{3}{4} = \frac{5}{2} \div \frac{3}{4} = \frac{5}{2} \times \frac{4}{3} = \frac{5 \times 4}{2 \times 3} = \frac{5 \times \overset{2}{\cancel{4}}}{\underset{1}{\cancel{2}} \times 3} = \frac{5 \times 2}{1 \times 3} = \frac{10}{3} = 3\frac{1}{3}$

42 $-8\frac{1}{3} \div 1\frac{1}{9} = -7\frac{1}{2}$

$$-8\frac{1}{3} \div 1\frac{1}{9} = -\frac{25}{3} \div \frac{10}{9} = -\frac{25}{3} \times \frac{9}{10} = -\frac{25 \times 9}{3 \times 10} = -\frac{\overset{5}{\cancel{25}} \times \overset{3}{\cancel{9}}}{\underset{1}{\cancel{3}} \times \underset{2}{\cancel{10}}} = -\frac{5 \times 3}{1 \times 2} = -\frac{15}{2} = -7\frac{1}{2}$$

43 $\frac{121}{36} \div \left(-\frac{22}{45}\right) = -6\frac{7}{8}$

$$\frac{121}{36} \div \left(-\frac{22}{45}\right) = \frac{121}{36} \times \left(-\frac{45}{22}\right) = -\frac{\overset{11}{\cancel{121}} \times \overset{5}{\cancel{45}}}{\underset{4}{\cancel{36}} \times \underset{2}{\cancel{22}}} = -\frac{11 \times 5}{4 \times 2} = -\frac{55}{8} = -6\frac{7}{8}$$

44 $7\frac{1}{7} \div 3\frac{3}{14} = 2\frac{2}{9}$

$$7\frac{1}{7} \div 3\frac{3}{14} = \frac{50}{7} \div \frac{45}{14} = \frac{50}{7} \times \frac{14}{45} = \frac{\overset{10}{\cancel{50}} \times \overset{2}{\cancel{14}}}{\underset{1}{\cancel{7}} \times \underset{9}{\cancel{45}}} = \frac{10 \times 2}{1 \times 9} = \frac{20}{9} = 2\frac{2}{9}$$

45 Change $\frac{3}{5}$ to a decimal. The answer is **0.6**. $\frac{3}{5} = \frac{6}{10} = 0.6$

46 Change $\frac{40}{9}$ to a decimal. The answer is $4.\overline{4}$.

$\frac{40}{9} = 4\frac{4}{9} = 4.\overline{4}$ because $9\overline{)4.00}$ gives $.44$, 36, 40, 36, 4, ...

47 Change $\frac{2}{11}$ to a decimal. The answer is $.\overline{18}$.

$\frac{2}{11} = .\overline{18}$ because $11\overline{)2.0000}$ gives $.1818$, 11, 90, 88, 20, 11, 90, 88, 2, ...

48 Change 0.45 to a fraction. The answer is $\frac{9}{20}$ because $0.45 = \frac{45}{100} = \frac{\overset{9}{\cancel{45}}}{\underset{20}{\cancel{100}}} = \frac{9}{20}$.

49 Change $0.\overline{36}$ to a fraction. The answer is $\frac{4}{11}$.

$0.\overline{36} = \frac{36}{99}$ because two digits repeat $= \frac{\overset{4}{\cancel{36}}}{\underset{11}{\cancel{99}}} = \frac{4}{11}$.

50 Change $0.\overline{405}$ to a fraction. The answer is $\frac{15}{37}$.

$0.\overline{405} = \frac{405}{999}$ because three digits repeat

$= \frac{\overset{45}{\cancel{405}}}{\underset{111}{\cancel{999}}} = \frac{\overset{15}{\cancel{45}}}{\underset{37}{\cancel{111}}} = \frac{15}{37}$.

Chapter 4

Discovering Exponents

In This Chapter
▶ Working with positive and negative exponents
▶ Using rules for operations with exponents
▶ Finding the power of powers
▶ Trying out scientific notation

*I*n the big picture of mathematics, exponents are a fairly new discovery. The principle behind them has always been there, but mathematicians had to first agree to use algebraic symbols such as x and y for values, and then they had to agree to the added shorthand of superscripts to indicate how many times the value was to be used (for example, instead of writing $xxxxx$, they agreed to write the x with an exponent of 5, such as x^5). In any case, be grateful. Exponents make life a lot easier.

This chapter provides you enough sample questions on exponents that you can introduce them to your parents as your new best friend.

Multiplying Numbers with Exponents

The number 16 can be written as 2^4 and the number 64 can be written as 2^6. When multiplying these two numbers together, you can either write $16 \times 64 = 1{,}024$ or you can multiply their two exponential forms together to get $2^4 \times 2^6 = 2^{10}$, which is equal to 1,024. The computation is easier — the numbers are smaller — when you use the exponential forms.

To multiply numbers with the same base (b), you add their exponents. The bases must be the same, or this rule doesn't work.

$$b^m \times b^n = b^{m+n}$$

Q. $3^5 \times 3^{-3} \times 7^{-2} \times 7^{-5} =$

A. $3^2 \times 7^{-7}$. The two factors with bases of 3 multiply as do the two with bases of 7, but they don't mix together. The negative exponents probably look intriguing. You can find an explanation of what they're all about in the "Using Negative Exponents" section later in this chapter.

So $3^5 \times 3^{-3} \times 7^{-2} \times 7^{-5} = 3^{5+(-3)} \times 7^{-2+(-5)} = 3^2 \times 7^{-7}$.

Q. $5^6 \times 5^{-7} \times 5 =$

A. **1.** In this case, the factor 5 is actually 5^1, because the exponent 1 usually isn't shown. Also, 5^0, or any non-zero number raised to the zero power is equal to 1.

So $5^6 \times 5^{-7} \times 5 = 5^{6+(-7)+1} = 5^0 = 1$.

48 Part I: Getting a Grip on Basic Concepts

1. $2^3 \times 2^4 =$

Solve It

2. $3^6 \times 3^{-4} =$

Solve It

3. $2^5 \times 2^{15} \times 3^4 \times 3^3 =$

Solve It

4. $7^{-3} \times 3^2 \times 5 \times 7^9 \times 5^4 =$

Solve It

Dividing Numbers with Exponents

When numbers appear in exponential form, you can divide them simply by subtracting their exponents. As with multiplication, the bases have to be the same in order to perform this operation.

When the bases are the same and two factors are divided, subtract their exponents:

$$\frac{b^m}{b^n} = b^{m-n}$$

Q. $\frac{3^4}{3^3} =$

A. **3.** $\frac{3^4}{3^3} = 3^{4-3} = 3^1 = 3$. Don't forget: When a number doesn't have an exponent showing, the exponent is 1.

Q. $\frac{8^2}{8^{-1}} =$

A. **512.** $\frac{8^2}{8^{-1}} = 8^{2-(-1)} = 8^3 = 512$. Don't forget: When subtracting signed numbers, change the problem to addition, and change the sign of that second number.

5. $\dfrac{3^4 \times 2^4}{3^2 \times 2^3} =$

Solve It

6. $\dfrac{3^2 \times 2^{-1}}{3 \times 2^{-5}} =$

Solve It

7. $\dfrac{7^{-3} \times 2^4 \times 5}{7^{-7} \times 2^4 \times 5^{-1}} =$

Solve It

8. $\dfrac{3^2 \times 3^4 \times 2}{3^{-1} \times 3^6 \times 2^{-3}} =$

Solve It

Raising Powers to Powers

Raising a power to a power means that you take a number in exponential form and raise it to some power. For instance, raising 3^6 to the fourth power means to multiply the sixth power of 3 by itself four times; in symbols, it looks like this: $(3^6)^4$. Raising something to a power tells you how many times it's multiplied by itself.

When raising a power to a power, don't forget these rules:

$(b^m)^n = b^{m \times n}$ So, to raise 3^6 to the fourth power, write $(3^6)^4 = 3^{6 \times 4} = 3^{24}$.

$(a \times b)^m = a^m \times b^m$ or $(a^p \times b^q)^m = a^{pm} \times b^{qm}$

$\left(\dfrac{a}{b}\right)^m = \dfrac{a^m}{b^m}$ or $\left(\dfrac{a^p}{b^q}\right)^m = \dfrac{a^{pm}}{b^{qm}}$

These rules say that if you multiply or divide two numbers and are raising the product or quotient to a power, then each factor gets raised to that power. (Remember, a *product* is the result of multiplying, and a *quotient* is the result of dividing.)

Part I: Getting a Grip on Basic Concepts

Q. $(3^{-4} \times 5^6)^7 =$

A. $3^{-28} \times 5^{42}$

$(3^{-4} \times 5^6)^7 = 3^{-4 \times 7} \times 5^{6 \times 7} = 3^{-28} \times 5^{42}$

Q. $\left(\dfrac{2^5}{5^2}\right)^3 =$

A. $\dfrac{2^{15}}{5^6}$

$\left(\dfrac{2^5}{5^2}\right)^3 = \dfrac{2^{5 \times 3}}{5^{2 \times 3}} = \dfrac{2^{15}}{5^6}$

---

9. $(3^2)^4 =$

Solve It

10. $(2^{-6})^{-8} =$

Solve It

---

11. $(2^3 \times 3^2)^4 =$

Solve It

12. $((3^5)^2)^6 =$

Solve It

---

13. $\left(\dfrac{2^2 \times 3^4}{5^2 \times 3}\right)^3 =$

Solve It

14. $\left(\dfrac{2^3}{7^5}\right)^2 =$

Solve It

Using Negative Exponents

Negative exponents are very useful in algebra because they allow you to do computations on numbers with the same base without having to deal with fractions.

The negative exponent is $b^{-n}=\frac{1}{b^n}$ and $\frac{1}{b^{-n}}=b^n$.

So, if you did problems in the sections earlier in this chapter and didn't like leaving all those negative exponents, you could have instead written them using fractions.

Another nice feature of negative exponents is how they affect fractions. Look at this rule:

$$\left(\frac{a^p}{b^q}\right)^{-n} = \left(\frac{b^q}{a^p}\right)^{n} = \frac{b^{qn}}{a^{pn}}$$

A quick, easy way of explaining this rule is to just say that a negative exponent flips the fraction and then applies a positive power to the factors.

Q. $4^3 \times 3^{-1} \times 6 \times 8^{-2} =$

A. 2. Move the factors with the negative exponents to the bottom, and their exponents then become positive. Then the fraction can be reduced.

$$4^3 \times 3^{-1} \times 6 \times 8^{-2} = \frac{4^3 \times 6^1}{3^1 \times 8^2} = \frac{64 \times 6^1}{3^1 \times 64} = \frac{6}{3} = 2$$

Q. $\left(\frac{3^4 \times 2^3}{3^7 \times 2^2}\right)^{-1} =$

A. $\frac{3^3}{2}$. Do you want another approach? Raise each of the powers in the parenthesis to the negative first power at any earlier step. The results are the same:

$$\left(\frac{3^4 \times 2^3}{3^7 \times 2^2}\right)^{-1} = \left(3^{-3} \times 2^1\right)^{-1} = \left(\frac{2^1}{3^3}\right)^{-1} = \left(\frac{3^3}{2^1}\right)^{1} = \frac{3^3}{2}$$

---

15. Rewrite $\frac{1}{3^6}$ using a negative exponent.

16. Rewrite $\frac{1}{5^{-5}}$ getting rid of the negative exponent.

Solve It

17. Simplify $\left(\dfrac{2^3}{3^2}\right)^{-1}$ leaving no negative exponent.

Solve It

18. Simplify $\left(\dfrac{3^{-4}}{2^3}\right)^{-2}$ leaving no negative exponent.

Solve It

19. Simplify $\dfrac{2^3 \times 3^{-4}}{2^{-2} \times 3^9}$ leaving no negative exponent.

Solve It

20. Simplify $\dfrac{\left(2^3 \times 3^2\right)^4}{\left(2^5 \times 3^{-1}\right)^5}$ leaving no negative exponent.

Solve It

Writing Numbers with Scientific Notation

Scientific notation is a standard way of writing numbers that are very small or very large in a more compact and useful way. If a scientist wants to talk about the distance to a star being 45,600,000,000,000,000,000,000,000 light years away, comparing this distance and working with it if it's written 4.56×10^{25} is easier.

A number written in scientific notation is the product of a number between 1 and 10 and a power of 10. The power tells how many decimal places the original decimal point was moved in order to make that first number be between 1 and 10. The power is negative if it's a very small number and positive if it's a very large number.

Q. 32,000,000,000 =

A. 3.2×10^{10}. Many modern scientific calculators show numbers in scientific notation with the letter E. So if you see 3.2 E 10 it means 3.2×10^{10} or 32,000,000,000.

Q. 0.00000000032 =

A. 3.2×10^{-10}

21. Write 4.03×10^{14} without scientific notation.

Solve It

22. Write 3.71×10^{-13} without scientific notation.

Solve It

23. Write 4,500,000,000,000,000,000 using scientific notation.

Solve It

24. Write 0.0000000000000003267 using scientific notation.

Solve It

Answers to Problems on Discovering Exponents

This section provides the answers (in bold) to the practice problems in this chapter.

1 $2^3 \times 2^4 = \mathbf{2^7}$ **or 128**

$2^3 \times 2^4 = 2^{3+4} = 2^7$

2 $3^6 \times 3^{-4} = \mathbf{3^2}$ **or 9**

$3^6 \times 3^{-4} = 3^{6+(-4)} = 3^2$

3 $2^5 \times 2^{15} \times 3^4 \times 3^3 = \mathbf{2^{20} \times 3^7}$

$2^5 \times 2^{15} \times 3^4 \times 3^3 = 2^{5+15} \times 3^{4+3} = 2^{20} \times 3^7$

4 $7^{-3} \times 3^2 \times 5 \times 7^9 \times 5^4 = \mathbf{7^6 \times 3^2 \times 5^5}$

$7^{-3} \times 3^2 \times 5 \times 7^9 \times 5^4$
$= 7^{-3} \times 7^9 \times 3^2 \times 5^1 \times 5^4$
$= 7^{-3+9} \times 3^2 \times 5^{1+4}$
$= 7^6 \times 3^2 \times 5^5$

5 $\dfrac{3^4 \times 2^4}{3^2 \times 2^3} = \mathbf{3^2 \times 2}$ **or 18**

$\dfrac{3^4 \times 2^4}{3^2 \times 2^3} = 3^{4-2} \times 2^{4-3} = 3^2 \times 2^1 = 3^2 \times 2$ or 18

6 $\dfrac{3^2 \times 2^{-1}}{3 \times 2^{-5}} = \mathbf{3 \times 2^4}$ **or 48**

$\dfrac{3^2 \times 2^{-1}}{3 \times 2^{-5}} = 3^{2-1} \times 2^{-1-(-5)} = 3^1 \times 2^4 = 3 \times 2^4$ or 48

7 $\dfrac{7^{-3} \times 2^4 \times 5}{7^{-7} \times 2^4 \times 5^{-1}} = \mathbf{7^4 \times 5^2}$

$\dfrac{7^{-3} \times 2^4 \times 5}{7^{-7} \times 2^4 \times 5^{-1}} = 7^{-3-(-7)} \times 2^{4-4} \times 5^{1-(-1)} = 7^4 \times 2^0 \times 5^2 = 7^4 \times 1 \times 5^2 = 7^4 \times 5^2$

8 $\dfrac{3^2 \times 3^4 \times 2}{3^{-1} \times 3^6 \times 2^{-3}} = \mathbf{3 \times 2^4}$ **or 48**

$\dfrac{3^2 \times 3^4 \times 2}{3^{-1} \times 3^6 \times 2^{-3}} = \left(3^{2+4-(-1)-6}\right) \times 2^{1-(-3)} = 3^1 \times 2^4 = 3 \times 2^4$ or 48

9 $\left(3^2\right)^4 = \mathbf{3^8}$

$\left(3^2\right)^4 = 3^{2 \times 4} = 3^8$

10 $\left(2^{-6}\right)^{-8} = \mathbf{2^{48}}$

$\left(2^{-6}\right)^{-8} = 2^{(-6) \times (-8)} = 2^{48}$

11 $\left(2^3 \times 3^2\right)^4 = \mathbf{2^{12} \times 3^8}$

$\left(2^3 \times 3^2\right)^4 = 2^{3 \times 4} \times 3^{2 \times 4} = 2^{12} \times 3^8$

Chapter 4: Discovering Exponents

12 $\left(\left(3^5\right)^2\right)^6 = 3^{60}$

$\left(\left(3^5\right)^2\right)^6 = \left(3^{5\times 2}\right)^6 = 3^{(5\times 2)6} = 3^{60}$

13 $\left(\dfrac{2^2 \times 3^4}{5^2 \times 3}\right)^3 = \dfrac{2^6 \times 3^9}{5^6}$

$\left(\dfrac{2^2 \times 3^4}{5^2 \times 3}\right)^3 = \left(\dfrac{2^2 \times 3^{4-1}}{5^2}\right)^3 = \left(\dfrac{2^2 \times 3^3}{5^2}\right)^3 = \dfrac{2^{2\times 3} \times 3^{3\times 3}}{5^{2\times 3}} = \dfrac{2^6 \times 3^9}{5^6}$

14 $\left(\dfrac{2^3}{7^5}\right)^2 = \dfrac{2^6}{7^{10}}$

$\left(\dfrac{2^3}{7^5}\right)^2 = \dfrac{2^{3(2)}}{7^{5(2)}} = \dfrac{2^6}{7^{10}}$

15 Rewrite $\dfrac{1}{3^6}$ using a negative exponent. The answer is 3^{-6}.

16 Rewrite $\dfrac{1}{5^{-5}}$ getting rid of the negative exponent. The answer is 5^5.

17 Simplify $\left(\dfrac{2^3}{3^2}\right)^{-1}$ leaving no negative exponent. The answer is $\dfrac{3^2}{2^3}$.

$\left(\dfrac{2^3}{3^2}\right)^{-1}\left(\dfrac{3^2}{2^3}\right)^1 = \dfrac{3^{2\times 1}}{2^{3\times 1}} = \dfrac{3^2}{2^3}$

18 Simplify $\left(\dfrac{3^{-4}}{2^3}\right)^{-2}$ leaving no negative exponent. The answer is $2^6 \times 3^8$.

$\left(\dfrac{3^{-4}}{2^3}\right)^{-2}\left(\dfrac{2^3}{3^{-4}}\right)^2 = \dfrac{2^{3\times 2}}{3^{-4\times 2}} = \dfrac{2^6}{3^{-8}} = 2^6 \times 3^8$ because $\dfrac{1}{3^{-8}} = 3^8$

19 Simplify $\dfrac{2^3 \times 3^{-4}}{2^{-2} \times 3^9}$ leaving no negative exponent. The answer is $\dfrac{2^5}{3^{13}}$.

$\dfrac{2^3 \times 3^{-4}}{2^{-2} \times 3^9} = 2^{3-(-2)} \times 3^{-4-9} = 2^5 \times 3^{-13} = \dfrac{2^5}{3^{13}}$

20 Simplify $\dfrac{\left(2^3 \times 3^2\right)^4}{\left(2^5 \times 3^{-1}\right)^5}$ leaving no negative exponent. The answer is $\dfrac{3^{13}}{2^{13}}$.

$\dfrac{\left(2^3 \times 3^2\right)^4}{\left(2^5 \times 3^{-1}\right)^5} = \dfrac{2^{3\times 4} \times 3^{2\times 4}}{2^{5\times 5} \times 3^{(-1)\times 5}} = \dfrac{2^{12} \times 3^8}{2^{25} \times 3^{-5}} = 2^{12-25} \times 3^{8-(-5)} = 2^{-13} \times 3^{13} = \dfrac{3^{13}}{2^{13}}$

21 Write 4.03×10^{14} without scientific notation. The answer is **403,000,000,000,000**. Move the decimal point 14 places to the right.

22 Write 3.71×10^{-13} without scientific notation. The answer is **0.000000000000371**. Move the decimal point 13 places to the left.

23 Write 4,500,000,000,000,000,000 using scientific notation. The answer is 4.5×10^{18}.

24 Write 0.0000000000000003267 using scientific notation. The answer is 3.267×10^{-16}.

Chapter 5

Taming the Radicals

In This Chapter
- Making expressions simpler with radicals
- Trimming down fractions
- Altering radicals to fractional exponents
- Performing operations on fractional exponents and radicals

*T*he operation of taking a square root, cube root, or any other root is an important one in algebra (as well as science and other areas of mathematics). The radical symbol ($\sqrt{}$) indicates that you want to take a *root* (what multiplies itself to give you the number or value) of an expression. A more convenient notation is to use a superscript, or power. This *superscript,* or exponent, is easily incorporated into algebraic work and makes computations easier to perform and express.

Simplifying Radical Expressions

Simplifying a radical expression means to rewrite it using the smallest possible numbers under the radical symbol. If the number under the radical isn't a *perfect square* or *cube* (a number that's the result of multiplying a single number times itself two times for a square, three times for a cube) of the root, then you want to see if there's a factor of that number that's a perfect square or cube and factor it out.

The square root of a product is equal to the product of two square roots containing the factors. The rule is $\sqrt{a \times b} = \sqrt{a} \times \sqrt{b}$. If the number under the radical is already a perfect square, then the problem is $\sqrt{25} = 5$ or $\sqrt{196} = 14$.

In the following two examples, the problems are written as the product of two factors — one of them is a perfect square factor. First, determine the square root and then write the expression in simplified form.

Part I: Getting a Grip on Basic Concepts

Q. $\sqrt{40} =$

A. $2\sqrt{10}$

$\sqrt{40} = \sqrt{4 \times 10} = \sqrt{4} \times \sqrt{10} = 2\sqrt{10}$

Q. $\sqrt{8} + \sqrt{18} =$

A. $5\sqrt{2}$. In this example, you can't add the two radicals together the way they are, but after you simplify them, the two terms have the same radical factor and you can add them together:

$\sqrt{8} + \sqrt{18} = \sqrt{4 \times 2} + \sqrt{9 \times 2}$
$= \sqrt{4} \times \sqrt{2} + \sqrt{9} \times \sqrt{2}$
$= 2\sqrt{2} + 3\sqrt{2} = 5\sqrt{2}$

1. Simplify $\sqrt{12}$.

Solve It

2. Simplify $\sqrt{200}$.

Solve It

3. Simplify $\sqrt{63}$.

Solve It

4. Simplify $\sqrt{75}$.

Solve It

5. Simplify the radicals in $\sqrt{24} + \sqrt{54}$ before adding.

Solve It

6. Simplify $\sqrt{72} - \sqrt{50}$ before subtracting.

Solve It

Rationalizing Fractions

You can rationalize a fraction with a radical in its denominator (bottom) by changing the original fraction to an equivalent fraction that has the radical in the numerator (top). Usually, you want to remove the radical from the denominator, because you'd be dividing by an irrational number. The square root of a number that isn't a perfect square is *irrational*. Dividing with an irrational number is difficult, because those numbers never end and never have a repeating pattern.

To rationalize a fraction with a square root in it, multiply both the numerator and denominator by that square root.

Q. $\dfrac{10}{\sqrt{5}}$

A. $2\sqrt{5}$. Use the property that $\sqrt{a} \times \sqrt{b} = \sqrt{a \times b}$, which works both ways. Doing so creates a perfect square in the denominator. Simplify that radical and then reduce the fraction:

$$\dfrac{10}{\sqrt{5}} = \dfrac{10}{\sqrt{5}} \times \dfrac{\sqrt{5}}{\sqrt{5}} = \dfrac{10\sqrt{5}}{\sqrt{25}}$$

$$= \dfrac{10\sqrt{5}}{5} = \dfrac{\overset{2}{\cancel{10}}\sqrt{5}}{\cancel{5}_1} = 2\sqrt{5}$$

Q. $\dfrac{8}{\sqrt{10}}$

A. $\dfrac{4\sqrt{10}}{5}$. In this problem, I multiply the numerator and denominator both by the radical in the denominator. The resulting fraction can be reduced:

$$\dfrac{8}{\sqrt{10}} = \dfrac{8}{\sqrt{10}} \times \dfrac{\sqrt{10}}{\sqrt{10}} = \dfrac{8\sqrt{10}}{\sqrt{100}} = \dfrac{8\sqrt{10}}{10}$$

$$= \dfrac{\overset{4}{\cancel{8}}\sqrt{10}}{\cancel{10}_5} = \dfrac{4\sqrt{10}}{5}$$

7. Rationalize $\dfrac{1}{\sqrt{2}}$.

Solve It

8. Rationalize $\dfrac{4}{\sqrt{3}}$.

Solve It

9. Rationalize $\dfrac{3}{\sqrt{6}}$.

Solve It

10. Rationalize $\dfrac{24}{\sqrt{15}}$.

Solve It

Changing Radicals to Exponents

The radical symbol indicates that you're to do the operation of taking a root — or figuring out what number multiplied itself to give you the value under the radical. An alternate notation, a *fractional exponent*, also indicates that you're to take a root, but it's much more efficient when combining factors when they all are powers of the same number.

The equivalence between root a and the fractional notation is $\sqrt{a} = a^{1/2}$. The general equivalence between all roots, powers, and fractional exponents is $\sqrt[n]{a^m} = a^{m/n}$.

Q. $\sqrt[3]{x^7} =$

A. $x^{7/3}$. The root is 3 — you're taking a cube root. The 3 goes in the fraction's denominator. The 7 goes in the fraction's numerator.

Q. $\dfrac{1}{\sqrt{x^3}} =$

A. $x^{-3/2}$. The exponent becomes negative when you bring up the factor from the fraction's denominator. Also, no root is showing on the radical, so it's assumed that a 2 goes there, and it's a square root.

11. Write the radical form in exponential form: $\sqrt{6}$

Solve It

12. Write the radical form in exponential form: $\sqrt[3]{x}$

Solve It

13. Write the radical form in exponential form: $\sqrt{7^5}$

Solve It

14. Write the radical form in exponential form: $\sqrt[4]{y^3}$

Solve It

15. Write the radical form in exponential form:

$$\frac{1}{\sqrt{x}}$$

Solve It

16. Write the radical form in exponential form:

$$\frac{3}{\sqrt[5]{2^2}}$$

Solve It

Using Fractional Exponents

Fractional exponents by themselves are fine and dandy. They're a nice, compact way of writing an operation to be performed on a factor. What's even nicer is when you can simplify the expression and get rid of the fractional exponent. Taking this step is the preferred form.

TIP If a value is written $a^{m/n}$, the easiest way to solve it is to take the root first and then raise the result to the power $\left(a^{1/n}\right)^m$. Doing so keeps the numbers relatively small — or at least smaller than the power might become. The answer comes out the same, either way. Being able to compute these in your head saves time.

Q. $8^{4/3}$

A. 16. This problem is easier than raising 8 to the fourth power, which is 4,096, and then taking the cube root of that big number. This way, you can do it all in your head. If you write out the solution, here's what it would look like:

$$8^{4/3} = \left(8^{1/3}\right)^4 = \left(\sqrt[3]{8}\right)^4 = 2^4 = 16$$

Q. $\left(\frac{1}{9}\right)^{3/2}$

A. $\frac{1}{27}$. Use the rule from Chapter 4 on raising a fraction to a power. When the number 1 is raised to any power, the result is always 1. The rest involves the denominator.

$$\left(\frac{1}{9}\right)^{3/2} = \frac{1^{3/2}}{9^{3/2}} = \frac{1}{\left(9^{1/2}\right)^3} = \frac{1}{\left(\sqrt{9}\right)^3} = \frac{1}{3^3} = \frac{1}{27}$$

17. Compute the value of $4^{5/2}$.

Solve It

18. Compute the value of $27^{2/3}$.

Solve It

Part I: Getting a Grip on Basic Concepts

19. Compute the value of $8^{5/3}$.

Solve It

20. Compute the value of $9^{3/2}$.

Solve It

21. Compute the value of $\left(\frac{1}{4}\right)^{3/2}$.

Solve It

22. Compute the value of $\left(\frac{8}{27}\right)^{4/3}$.

Solve It

Simplifying Expressions with Exponents

Having fractional exponents instead of radicals is better because fractional exponents are easier to work with in situations where a complicated or messy expression needs to be simplified into something neater. These situations usually involve factors with the same base that are multiplied and/or divided. When the bases are the same, the rules for multiplying (by adding exponents) and dividing (by subtracting exponents) and raising to powers (by multiplying exponents) are applied. Refer back to Chapter 4 if you need a reminder on these concepts. Here are some examples:

Q. $2^{4/3} \times 2^{5/3} =$

A. 8. Remember, when numbers with the same base are multiplied together, you add the exponents.

$$2^{4/3} \times 2^{5/3} = 2^{4/3 + 5/3} = 2^{9/3} = 2^3 = 8$$

Q. $\dfrac{5^{9/2}}{\sqrt{5}} =$

A. 625. First, change the square root of 5 to its exponential form. Then, because you're dividing numbers with the same base, subtract the exponents.

$$\frac{5^{9/2}}{\sqrt{5}} = \frac{5^{9/2}}{5^{1/2}} = 5^{9/2 - 1/2} = 5^{8/2} = 5^4 = 625$$

Chapter 5: Taming the Radicals

23. Simplify $2^{1/4} \times 2^{3/4}$.

Solve It

24. Simplify $\dfrac{6^{14/5}}{6^{4/5}}$.

Solve It

25. Simplify $8^{1/6} \sqrt{8}$.

Solve It

26. Simplify $\dfrac{4^{7/4}}{4^{1/4}}$.

Solve It

27. Simplify $3^{1/2} \times 3^{1/3} \times 3^{1/6}$.

Solve It

28. Simplify $\dfrac{9^{3/4} \times 9^{7/4}}{9}$.

Solve It

Answers to Problems on Radicals

This section provides the answers (in bold) to the practice problems in this chapter.

1. Simplify $\sqrt{12}$. The answer is **$2\sqrt{3}$**.

   $\sqrt{12} = \sqrt{4 \times 3} = \sqrt{4}\sqrt{3} = 2\sqrt{3}$

2. Simplify $\sqrt{200}$. The answer is **$10\sqrt{2}$**.

   $\sqrt{200} = \sqrt{100 \times 2} = \sqrt{100}\sqrt{2} = 10\sqrt{2}$

3. Simplify $\sqrt{63}$. The answer is **$3\sqrt{7}$**.

   $\sqrt{63} = \sqrt{9 \times 7} = \sqrt{9}\sqrt{7} = 3\sqrt{7}$

4. Simplify $\sqrt{75}$. The answer is **$5\sqrt{3}$**.

   $\sqrt{75} = \sqrt{25 \times 3} = \sqrt{25}\sqrt{3} = 5\sqrt{3}$

5. Simplify the radicals in $\sqrt{24} + \sqrt{54}$ before adding. The answer is **$5\sqrt{6}$**.

   $\sqrt{24} + \sqrt{54} = \sqrt{4 \times 6} + \sqrt{9 \times 6} = \sqrt{4}\sqrt{6} + \sqrt{9}\sqrt{6} = 2\sqrt{6} + 3\sqrt{6} = 5\sqrt{6}$

6. Simplify $\sqrt{72} - \sqrt{50}$ before subtracting. The answer is **$\sqrt{2}$**.

   $\sqrt{72} - \sqrt{50} = \sqrt{36 \times 2} - \sqrt{25 \times 2} = \sqrt{36}\sqrt{2} - \sqrt{25}\sqrt{2} = 6\sqrt{2} - 5\sqrt{2} = \sqrt{2}$

7. Rationalize $\dfrac{1}{\sqrt{2}}$. The answer is **$\dfrac{\sqrt{2}}{2}$**.

   $\dfrac{1}{\sqrt{2}} = \dfrac{1}{\sqrt{2}} \times \dfrac{\sqrt{2}}{\sqrt{2}} = \dfrac{\sqrt{2}}{\sqrt{4}} = \dfrac{\sqrt{2}}{2}$

8. Rationalize $\dfrac{4}{\sqrt{3}}$. The answer is **$\dfrac{4\sqrt{3}}{3}$**.

   $\dfrac{4}{\sqrt{3}} = \dfrac{4}{\sqrt{3}} \times \dfrac{\sqrt{3}}{\sqrt{3}} = \dfrac{4\sqrt{3}}{\sqrt{9}} = \dfrac{4\sqrt{3}}{3}$

9. Rationalize $\dfrac{3}{\sqrt{6}}$. The answer is **$\dfrac{\sqrt{6}}{2}$**.

   $\dfrac{3}{\sqrt{6}} = \dfrac{3}{\sqrt{6}} \times \dfrac{\sqrt{6}}{\sqrt{6}} = \dfrac{3\sqrt{6}}{\sqrt{36}} = \dfrac{\overset{1}{\cancel{3}}\sqrt{6}}{\underset{2}{\cancel{6}}} = \dfrac{\sqrt{6}}{2}$

10. Rationalize $\dfrac{24}{\sqrt{15}}$. The answer is **$\dfrac{8\sqrt{15}}{5}$**.

    $\dfrac{24}{\sqrt{15}} = \dfrac{24}{\sqrt{15}} \times \dfrac{\sqrt{15}}{\sqrt{15}} = \dfrac{24\sqrt{15}}{\sqrt{15 \times 15}} = \dfrac{24\sqrt{15}}{15} = \dfrac{\overset{8}{\cancel{24}}\sqrt{15}}{\underset{5}{\cancel{15}}} = \dfrac{8\sqrt{15}}{5}$

11. Write the radical form in exponential form: $\sqrt{6}$. The answer is **$6^{1/2}$**.

12. Write the radical form in exponential form: $\sqrt[3]{x}$. The answer is **$x^{1/3}$**.

13. Write the radical form in exponential form: $\sqrt{7^5}$. The answer is **$7^{5/2}$**.

14. Write the radical form in exponential form: $\sqrt[4]{y^3}$. The answer is **$y^{3/4}$**.

15. Write the radical form in exponential form: $\dfrac{1}{\sqrt{x}}$. The answer is **$x^{-1/2}$**.

16. Write the radical form in exponential form: $\dfrac{3}{\sqrt[5]{2^2}}$. The answer is **$3 \times 2^{-2/5}$**.

17. Compute the value of $4^{5/2}$. The answer is **32**.

    $4^{5/2} = \left(4^{1/2}\right)^5 = \left(\sqrt{4}\right)^5 = 2^5 = 32$

18 Compute the value of $27^{2/3}$. The answer is **9**.

$$27^{2/3} = \left(27^{1/3}\right)^2 = \left(\sqrt[3]{27}\right)^2 = 3^2 = 9$$

19 Compute the value of $8^{5/3}$. The answer is **32**.

$$8^{5/3} = \left(8^{1/3}\right)^5 = \left(\sqrt[3]{8}\right)^5 = 2^5 = 32$$

20 Compute the value of $9^{3/2}$. The answer is **27**.

$$9^{3/2} = \left(9^{1/2}\right)^3 = \left(\sqrt{9}\right)^3 = 3^3 = 27$$

21 Compute the value of $\left(\frac{1}{4}\right)^{3/2}$. The answer is **⅛**.

$$\left(\frac{1}{4}\right)^{3/2} = \frac{1^{3/2}}{4^{3/2}} = \frac{1}{\left(4^{1/2}\right)^3} = \frac{1}{2^3} = \frac{1}{8}$$

22 Compute the value of $\left(\frac{8}{27}\right)^{4/3}$. The answer is **16/81**.

$$\left(\frac{8}{27}\right)^{4/3} = \frac{8^{4/3}}{27^{4/3}} = \frac{\left(8^{1/3}\right)^4}{\left(27^{1/3}\right)^4} = \frac{2^4}{3^4} = \frac{16}{81}$$

23 Simplify $2^{1/4} \times 2^{3/4}$. The answer is **2**.

$$2^{1/4} \times 2^{3/4} = 2^{1/4 + 3/4} = 2^{4/4} = 2^1 = 2$$

24 Simplify $\frac{6^{14/5}}{6^{4/5}}$. The answer is **36**.

$$\frac{6^{14/5}}{6^{4/5}} = 6^{14/5 - 4/5} = 6^{10/5} = 6^2 = 36$$

25 Simplify $8^{1/6}\sqrt{8}$. The answer is **4**.

$$8^{1/6}\sqrt{8} = 8^{1/6} \times 8^{1/2} = 8^{1/6 + 1/2} = 8^{1/6 + 3/6} = 8^{4/6} = 8^{2/3} = \left(8^{1/3}\right)^2 = 2^2 = 4$$

26 Simplify $\frac{4^{7/4}}{4^{1/4}}$. The answer is **8**.

$$\frac{4^{7/4}}{4^{1/4}} = 4^{7/4 - 1/4} = 4^{6/4} = 4^{3/2} = \left(4^{1/2}\right)^3 = 2^3 = 8$$

27 Simplify $3^{1/2} \times 3^{1/3} \times 3^{1/6}$. The answer is **3**.

$$3^{1/2} \times 3^{1/3} \times 3^{1/6} = 3^{1/2 + 1/3 + 1/6} = 3^{3/6 + 2/6 + 1/6} = 3^{6/6} = 3^1 = 3$$

28 Simplify $\frac{9^{3/4} \times 9^{7/4}}{9}$. The answer is **27**.

$$\frac{9^{3/4} \times 9^{7/4}}{9} = \frac{9^{3/4 + 7/4}}{9} = \frac{9^{10/4}}{9} = \frac{9^{5/2}}{9} = 9^{5/2 - 1} = 9^{5/2 - 2/2} = 9^{3/2} = \left(9^{1/2}\right)^3 = 3^3 = 27$$

Chapter 6
Keeping It Simple for Algebraic Expressions

In This Chapter
- Operating on algebraic expressions
- Understanding order of operations
- Combining all the rules

The operations of addition, subtraction, multiplication, and division are familiar to grade school students. And the nice thing is that these operations work the same, no matter what kind of math you do. What happens here is that algebra introduces some twists to those rules. You can apply all these rules to the letters, which stand for variables in algebra. However, because you usually don't know what values the variables represent, you have to be careful when doing the operations and reporting the results.

This chapter offers you multitudes of problems to make sure you keep everything in order (unlike your bedroom, which your mother told me needed cleaned).

Adding and Subtracting Like Terms

In algebra, the expression "like terms" refers to when one or more variables are multiplied, divided, or both, sometimes with a number (*coefficient*) multiplied to tell you how many of that term you have (for example, $2xy$ means that you have two terms that are made up of an x). When adding and subtracting algebraic terms, the terms have to be *alike* as far as having the same variables raised to exactly the same power. For example, two terms that are *alike* are $2a^3b$ and $5a^3b$. Two terms that aren't *alike* are $3xyz$ and $4x^2yz$. Notice that the power on the x term is different in the two terms.

Q. $6a + 2b - 4a + 7b + 5 =$

A. $2a + 9b + 5$. First, change the order and group the like terms together, then compute:

$6a + 2b - 4a + 7b + 5 =$
$(6a - 4a) + (2b + 7b) + 5$

The parentheses aren't necessary, but they help to keep track of what you can combine.

Q. $8x^2 - 3x + 4xy - 9x^2 - 5x - 20xy =$

A. $-x^2 - 8x - 16xy$. Again, combine like terms and compute:

$8x^2 - 3x + 4xy - 9x^2 - 5x - 20xy =$
$(8x^2 - 9x^2) + (-3x - 5x) + (4xy - 20xy) =$
$-x^2 - 8x - 16xy$

Part I: Getting a Grip on Basic Concepts

1. Combine the like terms in
$4a + 3ab - 2ab + 6a$.

Solve It

2. Combine the like terms in
$3x^2y - 2xy^2 + 4x^3 - 8x^2y$.

Solve It

3. Combine the like terms in
$2a^2 + 3a - 4 + 7a^2 - 6a + 5$.

Solve It

4. Combine the like terms in
$ab + bc + cd + de - ab + 2bc + e$.

Solve It

Multiplying and Dividing Algebraic Factors

Multiplying and dividing algebraic expressions is somewhat different than adding and subtracting them. When multiplying and dividing, the terms don't have to be exactly alike. You can multiply or divide all variables with the same base — using the laws of exponents (check out Chapter 4 for more information) — and you multiply or divide the number factors.

If a variable's power is greater in the denominator, then the difference between the two powers is written as a positive power of the base — in the denominator. For example:

$$\frac{18a^3b^{12}}{3a^7b^4} = \frac{6b^{12-4}}{a^{7-3}} = \frac{6b^8}{a^4}$$

Q. $(4x^2y^2z^3)(3xy^4z^3) =$

A. $12x^3y^6z^6$. The product of 4 and 3 is 12. Multiply the x's to get $x^2 \times x = x^3$. Multiply the y's and z's and you get $y^2 \times y^4 = y^6$ and $z^3 \times z^3 = z^6$. Each variable has its own power determined by the factors multiplied together to get it.

Q. $\dfrac{32x^4y^3}{4x^2y^2}$

A. $8x^2y$. All the powers of the variables in the numerator are greater than those in the denominator, so you can write the result without a fraction. Here are the details: $\dfrac{32x^4y^3}{4x^2y^2} = \dfrac{8x^{4-2}y^{3-2}}{1} = \dfrac{8x^2y^1}{1} = 8x^2y$

Chapter 6: Keeping It Simple for Algebraic Expressions 69

5. Multiply $(3x)(2x^2)$.

Solve It

6. Multiply $(4y^2)(x^4y)$.

Solve It

7. Multiply $(6x^3y^2z^2)(8x^3y^4z)$.

Solve It

8. Divide (write all exponents as positive numbers) $\frac{10x^2y^3}{5xy^2}$.

Solve It

9. Divide (write all exponents as positive numbers) $\frac{24x}{3x^4}$.

Solve It

10. Divide (write all exponents as positive numbers) $\frac{13x^3y^4}{26x^8y^3}$.

Solve It

Using Order of Operation

Because much of algebra involves symbols for numbers, operations, and relationships and groupings, are you really surprised that *order* is something special, too? Order helps you solve problems by making sure everyone follows the same procedure and handling of expressions. Order is how mathematicians have been able to communicate — agreeing on these same conventions. The *order of operations* falls in to this category.

When performing operations on algebraic expressions and you have a choice between one or more operations to perform, use the following order:

1. **Perform all powers or roots, moving left to right.**
2. **Perform all multiplication or division, moving left to right.**
3. **Perform all addition or subtraction, moving left to right.**

70 Part I: Getting a Grip on Basic Concepts

These rules are *interrupted* if the problem has grouping symbols. You first need to perform operations in grouping symbols, such as ( ), { }, or [ ].

Q. $2 \times 4 - 10/5 =$

A. 6. First, do the multiplication and division and then the subtraction so

$2 \times 4 - 10/5 = 8 - 2 = 6$

Q. $\dfrac{8 + 2^2 \times 5}{2^3 - 1} =$

A. 4. First, find the two powers ($2^2 = 4$ and $2^3 = 8$). Then multiply in the numerator and subtract in the denominator. After you add the two terms in the numerator, then you can perform the final division. The fraction line acts as a grouping symbol — what's in the numerator and denominator have to be computed first. Here's how it breaks down:

$$\dfrac{8 + 2^2 \times 5}{2^3 - 1} = \dfrac{8 + 4 \times 5}{8 - 1} = \dfrac{8 + 20}{8 - 1} = \dfrac{28}{7} = 4$$

11. $5 + 3 \times 4^2 =$

Solve It

12. $6 \div 2 - 5\sqrt{9} =$

Solve It

13. $2 \times 3^3 + 3(2^2 - 5) =$

Solve It

14. $\dfrac{4^2 + 3^2}{9(4) - 11} =$

Solve It

Evaluating Expressions with Order of Operations

Evaluating an expression means that you want to change it from a bunch of letters and numbers to a specific value — some number. After you solve an equation or *inequality* (an expression with > or < in it), you want to go back and check to see if your solution really works — which is called evaluating the expression. For instance, if you let $x = 2$ in the expression $3x^2 - 2x + 1$, you replace all the x's with 2s and apply the order of operations when performing the different operations involved. In this case, you get $3(2)^2 - 2(2) + 1 = 3(4) - 4 + 1 = 12 - 4 + 1 = 9$. Can you see why knowing that you square the 2 before multiplying by the 3 is so important? If you multiply by the 3, first, you'd end up with that first term being worth 36 instead of 12. It makes a big difference.

Q. Evaluate $\frac{5y - y^2}{2x}$ when $y = -4$ and $x = -3$.

A. 6

$$\frac{5y - y^2}{2x} = \frac{5(-4) - (-4)^2}{2(-3)} = \frac{5(-4) - 16}{2(-3)} = \frac{-20 - 16}{-6} = \frac{-36}{-6} = 6$$

Q. What's the Spanish word for belly button?

A. **Ombligo.** Didn't know you were going to get a little cultural diversity in an algebra book, did you? Now go clean out your ombligo.

15. Evaluate $3x^2$ if $x = -2$.

Solve It

16. Evaluate $9y - y^2$ if $y = -1$.

Solve It

17. Evaluate $3x - 2y$ if $x = 4$ and $y = 3$.

Solve It

18. Evaluate $6x^2 - xy$ if $x = 2$ and $y = -3$.

Solve It

19. Evaluate $\frac{2x+y}{x-y}$ if $x = 4$ and $y = 1$.

Solve It

20. Evaluate $\frac{x^2 - 2x}{y^2 + 2y}$ if $x = 3$ and $y = -1$.

Solve It

Answers to Problems on Algebraic Expressions

This section provides the answers (in bold) to the practice problems in this chapter.

1 Combine the like terms in $4a + 3ab - 2ab + 6a$. The answer is **$10a + ab$**.

$4a + 3ab - 2ab + 6a = (4a + 6a) + (3ab - 2ab) = 10a + ab$

2 Combine the like terms in $3x^2y - 2xy^2 + 4x^3 - 8x^2y$. The answer is **$-5x^2y - 2xy^2 + 4x^3$**.

$3x^2y - 2xy^2 + 4x^3 - 8x^2y = (3x^2y - 8x^2y) - 2xy^2 + 4x^3 = -5x^2y - 2xy^2 + 4x^3$

3 Combine the like terms in $2a^2 + 3a - 4 + 7a^2 - 6a + 5$. The answer is **$9a^2 - 3a + 1$**.

$2a^2 + 3a - 4 + 7a^2 - 6a + 5 = (2a^2 + 7a^2) + (3a - 6a) + (-4 + 5) = 9a^2 - 3a + 1$.

4 Combine the like terms in $ab + bc + cd + de - ab + 2bc + e$. The answer is **$3bc + cd + de + e$**.

$ab + bc + cd + de - ab + 2bc + e = (ab - ab) + (bc + 2bc) + cd + de + e = 0 + 3bc + cd + de + e$.

5 Multiply $(3x)(2x^2)$. The answer is **$6x^3$**. $(3x)(2x^2) = (3 \times 2)(x \times x^2) = 6x^3$

6 Multiply $(4y^2)(x^4y)$. The answer is **$4x^4y^3$**. $(4y^2)(x^4y) = 4x^4(y^2 \times y) = 4x^4y^3$

7 Multiply $(6x^3y^2z^2)(8x^3y^4z)$. The answer is **$48x^6y^6z^3$**.

$(6x^3y^2z^2)(8x^3y^4z) = (6 \times 8)(x^3 \times x^3)(y^2 \times y^4)(z^2 \times z) = 48x^6y^6z^3$

8 Divide (write all exponents as positive numbers) $\dfrac{10x^2y^3}{5xy^2}$. The answer is **$2xy$**.

$\dfrac{10x^2y^3}{5xy^2} = \dfrac{2x^{2-1}y^{3-2}}{1} = 2xy$

9 Divide (write all exponents as positive numbers) $\dfrac{24x}{3x^4}$. The answer is **$\dfrac{8}{x^3}$**.

$\dfrac{24x}{3x^4} = \dfrac{8}{x^{4-1}} = \dfrac{8}{x^3}$

10 Divide (write all exponents as positive numbers) $\dfrac{13x^3y^4}{26x^8y^3}$. The answer is **$\dfrac{y}{2x^5}$**.

$\dfrac{13x^3y^4}{26x^8y^3} = \dfrac{y^{4-3}}{2x^{8-3}} = \dfrac{y^1}{2x^5} = \dfrac{y}{2x^5}$

11 $5 + 3 \times 4^2 = $ **53**

$5 + 3 \times 4^2 = 5 + 3 \times 16 = 5 + 48 = 53$

12 $6 \div 2 - 5\sqrt{9} = $ **-12**

$6 \div 2 - 5\sqrt{9} = 6 \div 2 - 5 \times 3 = 3 - 15 = -12$

13 $2 \times 3^3 + 3(2^2 - 5) = $ **51**

$2 \times 3^3 + 3(2^2 - 5) = 2 \times 27 + 3(4 - 5) = 2 \times 27 + 3(-1) = 54 - 3 = 51$

14 $\dfrac{4^2 + 3^2}{9(4) - 11} = $ **1** (simple as that).

$\dfrac{4^2 + 3^2}{9(4) - 11} = \dfrac{16 + 9}{9(4) - 11} = \dfrac{16 + 9}{36 - 11} = \dfrac{25}{25} = 1$

15 Evaluate $3x^2$ if $x = -2$. The answer is **12**.

$3x^2 = 3(-2)^2 = 3 \times 4 = 12$

16 Evaluate $9y - y^2$ if $y = -1$. The answer is **-10**.

$9y - y^2 = 9(-1) - (-1)^2 = -9 - 1 = -10$

17 Evaluate $3x - 2y$ if $x = 4$ and $y = 3$. The answer is **6**.
$3x - 2y = 3(4) - 2(3) = 12 - 6 = 6$

18 Evaluate $6x^2 - xy$ if $x = 2$ and $y = -3$. The answer is **30**.
$6x^2 - xy = 6(2^2) - 2(-3) = 6(4) - 2(-3) = 24 + 6 = 30$

19 Evaluate $\dfrac{2x + y}{x - y}$ if $x = 4$ and $y = 1$. The answer is **3**.
$\dfrac{2x + y}{x - y} = \dfrac{2 \times 4 + 1}{4 - 1} = \dfrac{8 + 1}{4 - 1} = \dfrac{9}{3} = 3$

20 Evaluate $\dfrac{x^2 - 2x}{y^2 + 2y}$ if $x = 3$ and $y = -1$. The answer is **-3**.
$\dfrac{x^2 - 2x}{y^2 + 2y} = \dfrac{3^2 - 2 \times 3}{(-1)^2 + 2(-1)} = \dfrac{9 - 6}{1 - 2} = \dfrac{3}{-1} = -3$

Part II
Operating and Factoring

In This Part...

The *operating* portion of this part does include some cutting apart and stitching back together. You have to know where everything goes and how it gets there. *Factoring* is a word that either does it for you or doesn't. In algebra, factoring means something very specific. To factor means to break something down into its component parts so that it can be used for something else. The factored form is very useful — you just have to know how to get there. Conquer these skills, and you're on your way — able to leap tall built-up fractions, more powerful than a loco equation, and all this within your single hand. Go for it!

Chapter 7
Using Special Rules for Multiplying Expressions

In This Chapter
- Distributing factors and FOILing the enemy
- Squaring and cubing binomials
- Raising binomials to many powers
- Incorporating special rules for multiplying

In Chapter 6, I cover the basics of multiplying algebraic expressions. In this chapter, I discuss multiplying one or more terms with two or more terms. When doing so, you first need to understand some unique situations. They're special (just like you are, at least that's what your mother told you, right?), and they pop up so frequently that they deserve individual attention upfront.

You also have the nitty-gritty of multiplying factor by factor and term by term (the long way) and the neat little trick called FOIL. The big picture, though, is that you need to be able to multiply these different types of expressions. You need to know how to solve the problems so you can then later go backwards and factor them back to their original multiplication form.

Distributing One Factor over Many

When you *distribute* literature, you give one copy to each person. When you *distribute* some factor over several terms, it means that you take each factor (multiplier) on the outside of a grouping symbol and multiply it times each term (all separated by + or –) inside the grouping symbol.

The distributive rule is

$$a(b + c) = ab + ac \text{ and } a(b - c) = ab - ac$$

In other words, if you multiply a value times a sum or difference within a grouping symbol such as $2(x + 3)$, you multiply every term inside by the factor outside. In this case, multiply the x and 3 each by 2.

78 Part II: Operating and Factoring

Q. $6x(3x^2 + 5x - 2) =$

A. $18x^3 + 30x^2 - 12x$. Now for the filling in the middle: $6x(3x^2 + 5x - 2) = 6x(3x^2) + 6x(5x) - 6x(2) = 18x^3 + 30x^2 - 12x$

Q. What's the average temperature for the Bahamas in December?

A. **78° Fahrenheit**. Can you just imagine? Instead of a cold December day, you can relax on the beach, let the sun warm your body, feel the sand between your toes, and not have to think about algebra one bit. Okay now, wake up. Your dream is over.

1. Distribute $3(2x + 3y - 4z + 1)$.

Solve It

2. Distribute $x(8x^3 - 3x^2 + 2x - 5)$.

Solve It

3. Distribute $x^2y(2xy^2 + 3xyz + y^2z^3)$.

Solve It

4. Distribute $-4y(3y^4 - 2y^2 + 5y - 5)$.

Solve It

Getting FOILed Again

A common process in algebra is to multiply two binomials together. A *binomial* is an expression with two terms such as $x + 7$. One possible way to multiply the binomials together is to distribute the two terms over the two terms.

But some math whiz came up with a great acronym *FOIL*, which translates to F for First, O for Outer, I for Inner, and L for Last. This acronym helps you save time and makes solving binomials easier. These letters refer to the terms' position in the product of two binomials.

Chapter 7: Using Special Rules for Multiplying Expressions

FOIL the product $(a + b)(c + d)$

The product of the First terms is ac.

The product of the Outer terms is ad.

The product of the Inner terms is bc.

The product of the Last terms is bd.

The result is then $ac + ad + bc + bd$. Usually ad and bc are like terms and can be combined.

Q. Use FOIL to multiply $(x - 8)(x - 9)$.

A. $x^2 - 17x + 72$. Using FOIL, multiply the First, the Outer, the Inner, and the Last and then combine the like terms: $(x - 8)(x - 9) = x^2 - 9x - 8x + 72 = x^2 - 17x + 72$

Q. Use FOIL to multiply $(2x + 3)(x - 4)$.

A. $2x^2 - 5x - 12$. Using FOIL, multiply the First, the Outer, the Inner, and the Last and then combine the like terms: $(2x + 3)(x - 4) = 2x^2 - 8x + 3x - 12 = 2x^2 - 5x - 12$

5. Use FOIL to multiply $(2x + 1)(3x - 2)$.

Solve It

6. Use FOIL to multiply $(x - 7)(3x + 5)$.

Solve It

7. Use FOIL to multiply $(x^2 - 2)(x^2 - 4)$.

Solve It

8. Use FOIL to multiply $(3x + 4y)(4x - 3y)$.

Solve It

Squaring Binomials

You can always use FOIL or the distributive law to square a binomial, but you can also use a helpful pattern to make the work quick and easy.

The squares of binomials are

$$(a + b)^2 = a^2 + 2ab + b^2 \text{ and } (a - b)^2 = a^2 - 2ab + b^2$$

Q. $(x + 5)^2 =$

A. $x^2 + 10x + 25$. The $10x$ is twice the product of the x and 5: $(x + 5)^2 = (x)^2 + 2(x)(5) + (5)^2 = x^2 + 10x + 25$

Q. $(3y - 7)^2 =$

A. $9y^2 - 42y + 49$. The $42y$ is twice the product of $3y$ and 7. And, because the 7 is negative, the Inner and Outer products are, too: $(3y - 7)^2 = (3y)^2 - 2(3y)(7) + (7)^2 = 9y^2 - 42y + 49$

9. $(x + 3)^2 =$

Solve It

10. $(2y - 1)^2 =$

Solve It

11. $(3a - 2b)^2 =$

Solve It

12. $(5xy + z)^2 =$

Solve It

Multiplying the Sum and Difference of the Same Two Terms

When you multiply two binomials together, you can always just FOIL them. You save yourself some work, though, if you recognize that the terms in the two binomials are the same — except for the sign between them. If they're the sum and difference of the same two numbers, then their product is just the difference between the squares of the two terms.

The product of $(a + b)(a - b)$ is $a^2 - b^2$.

Q. $(x + 5)(x - 5) =$

A. $x^2 - 25$

Q. $(3ab - 4)(3ab + 4) =$

A. $9a^2b^2 - 16$

13. $(x + 3)(x - 3) =$

Solve It

14. $(2x - 7)(2x + 7) =$

Solve It

15. $(a^3 - 3)(a^2 + 3) =$

Solve It

16. $(2x^2h + 9)(2x^2h - 9) =$

Solve It

Cubing Binomials

To *cube* something in algebra is to multiply it by itself three times. In the case of cubing a binomial, you have a couple options.

First, you could square it and then multiply the original binomial times the square. You can do this by distributing and combining the like terms. Not a bad idea, but I have a better one.

When two binomials are cubed, patterns occur. The *coefficients* (numbers in front of and multiplying each term) in the answer are 1-3-3-1. The first coefficient is 1, the second is 3, the third is 3, and the last is 1. The other pattern is that the powers on the terms decrease and increase by ones. The powers of the first term in the binomial go down by one with each step, and the powers of the second term go up by one each time.

To cube a binomial, follow this rule:

The cube of $(a + b)^3$ is $a^3 + 3a^2b + 3ab^2 + b^3$.

Q. $(y + 4)^3 =$

A. $y^3 + 12y^2 + 48y + 64$. The answer is built by incorporating the two patterns — the 1-3-3-1 of the coefficients and the changes in the powers. $(y + 4)^3 = y^3 + 3y^2 \times 4 + 3y \times 4^2 + 4^3 = y^3 + 12y^2 + 48y + 64$. Notice how the powers of y go down one step each term. Also, the powers of 4, starting with the second term, go up one step each time. The last part of using this method is to simplify each term. The 1-3-3-1 pattern gets lost when you do the simplification, but it's still part of the answer — just hidden. *Note:* In binomials with a subtraction in them, the terms in the answer will have alternating signs: +, –, +, –.

Q. What's the state bird of Alaska?

A. **Willow Grouse**. Huh? If you guessed the penguin, guess again. If you live in the lower 48 states (or Hawaii), you may never see one of these rare beautiful birds. (Aren't you glad you got a nature lesson in an algebra book?)

17. $(x + 1)^3 =$

Solve It

18. $(y - 2)^3 =$

Solve It

19. $(3z + 1)^3 =$

Solve It

20. $(5 - 2y)^3 =$

Solve It

Creating the Sum and Difference of Cubes

A lot of what you do in algebra is to take advantage of patterns, rules, and quick tricks. Multiply the sum and difference of two values together, and you get the difference of squares. Another pattern gives you the sum or difference of two cubes. These patterns are actually going to mean a lot more to you when you do the factoring of binomials. For now, just practice with these patterns.

With this rule if you multiply a binomial with a particular *trinomial* (one that has the squares of the two terms in the binomial as well as the opposite of the product of them such as $(y - 5)(y^2 + 5y + 25)$, you get the sum or difference of two perfect cubes.

$$(a + b)(a^2 - ab + b^2) = a^3 + b^3 \text{ and } (a - b)(a^2 + ab + b^2) = a^3 - b^3$$

Q. $(y + 3)(y^2 - 3y + 9) =$

A. $y^3 + 27$. If you don't believe me, just multiply it out by distributing the binomial over the trinomial and combining like terms. You find all the *middle* terms pairing up with their opposites and becoming 0, leaving just the two cubes.

Q. $(5y - 1)(25y^2 + 5y + 1) =$

A. $125y^3 - 1$. The number 5 cubed is 125, and –1 cubed is –1. The other terms in the product will drop out because of the opposites that will appear there.

21. $(x - 2)(x^2 + 2x + 4) =$

Solve It

22. $(y + 1)(y^2 - y + 1) =$

Solve It

23. $(2z + 5)(4z^2 - 10z + 25) =$

Solve It

24. $(3x - 2)(9x^2 + 6x + 4) =$

Solve It

Raising Binomials to Higher Powers

The nice pattern for cubing binomials, 1-3-3-1, tells you what the coefficients of the different terms are — at least what they start out to be before simplifying the terms. Patterns also exist for raising binomials to the fourth power, fifth power, and so on. They're all based on mathematical *combinations* and are easily pulled out with the Pascal's Triangle. Check out Figure 7-1 to see a small piece of Pascal's Triangle with the powers of the binomial identified.

Figure 7-1: Pascal's Triangle can help you find powers of binomials.

```
            1              (a + b)⁰
          1   1            (a + b)¹
        1   2   1          (a + b)²
      1   3   3   1        (a + b)³
    1   4   6   4   1      (a + b)⁴
  1   5   10  10  5   1    (a + b)⁵
```

Q. Refer to Figure 7-1 and use the coefficients from the row for the fourth power of the binomial to raise $(x - 3y)^4$ to the fourth power.

A. $x^4 - 12x^3y + 54x^2y^2 - 108xy^3 + 81y^4$

$(x - 3y)^4 = (x + (-3y))^4$

Insert the coefficients 1-4-6-4-1 from the row in Pascal's triangle. (Actually, the ones are understood.)

$= x^4 + 4 \times x^3(-3y)^1 + 6 \times x^2(-3y)^2 + 4 \times x^1(-3y)^3 + (-3y)^4$

$= x^4 + 4 \times x^3(-3y) + 6 \times x^2(9y^2) + 4 \times x(-27y^3) + (81y^4)$

$= x^4 - 12x^3y + 54x^2y^2 - 108xy^3 + 81y^4$

Q. Raise $(2x + 5y)^3$ to the third power.

A. $8x^3 + 60x^2y + 150xy^2 + 125y^3$

Insert the coefficients 1-3-3-1 from the row in Pascal's triangle. (Actually, the ones are understood.)

$= (2x)^3 + 3 \times (2x)^2(5y)^1 + 3 \times (2x)^1(5y)^2 + (5y)^3$

$= 8x^3 + 3 \times (4x^2)(5y) + 3 \times (2x)(25y^2) + 125y^3$

$= 8x^3 + 60x^2y + 150xy^2 + 125y^3$

Notice how in the first step, the first term has decreasing powers of the exponent, and the second term has increasing powers. The last step has alternating signs. This method may seem rather complicated, but it still beats multiplying it out the long way.

25. $(x+1)^4 =$

Solve It

26. $(2y-1)^4 =$

Solve It

27. $(z-1)^5 =$

Solve It

28. $(3z+2)^5 =$

Solve It

Answers to Problems on Multiplying Expressions

This section provides the answers (in bold) to the practice problems in this chapter.

1. Distribute $3(2x + 3y - 4z + 1)$. The answer is **$6x + 9y - 12z + 3$**.
$$3(2x + 3y - 4z + 1) = 3(2x) + 3(3y) - 3(4z) + 3(1)$$
$$= 6x + 9y - 12z + 3$$

2. Distribute $x(8x^3 - 3x^2 + 2x - 5)$. The answer is **$8x^4 - 3x^3 + 2x^2 - 5x$**.
$$x(8x^3 - 3x^2 + 2x - 5) = x(8x^3) - x(3x^2) + x(2x) - x(5)$$
$$= 8x^4 - 3x^3 + 2x^2 - 5x$$

3. Distribute $x^2y(2xy^2 + 3xyz + y^2z^3)$. The answer is **$2x^3y^3 + 3x^3y^2z + x^2y^3z^3$**.
$$x^2y(2xy^2 + 3xyz + y^2z^3) = x^2y(2xy^2) + x^2y(3xyz) + x^2y(y^2z^3)$$
$$= 2x^3y^3 + 3x^3y^2z + x^2y^3z^3$$

4. Distribute $-4y(3y^4 - 2y^2 + 5y - 5)$. The answer is **$-12y^5 + 8y^3 - 20y^2 + 20y$**.
$$-4y(3y^4 - 2y^2 + 5y - 5) = (-4y)(3y^4 - 2y^2 + 5y - 5)$$
$$= (-4y)(3y^4) - (-4y)(2y^2) + (-4y)(5y) - (-4y)(5)$$
$$= -12y^5 + 8y^3 - 20y^2 + 20y$$

5. Use FOIL to multiply $(2x + 1)(3x - 2)$. The answer is **$6x^2 - x - 2$**.
$$(2x + 1)(3x - 2) = 6x^2 - 4x + 3x - 2$$
$$= 6x^2 - x - 2$$

6. Use FOIL to multiply $(x - 7)(3x + 5)$. The answer is **$3x^2 - 16x - 35$**.
$$(x - 7)(3x + 5) = 3x^2 + 5x - 21x - 35$$
$$= 3x^2 - 16x - 35$$

7. Use FOIL to multiply $(x^2 - 2)(x^2 - 4)$. The answer is **$x^4 - 6x^2 + 8$**.
$$(x^2 - 2)(x^2 - 4) = x^4 - 4x^2 - 2x^2 + 8$$
$$= x^4 - 6x^2 + 8$$

8. Use FOIL to multiply $(3x + 4y)(4x - 3y)$. The answer is **$12x^2 + 7xy - 12y^2$**.
$$(3x + 4y)(4x - 3y) = 12x^2 - 9xy + 16xy - 12y^2$$
$$= 12x^2 + 7xy - 12y^2$$

9. $(x + 3)^2 =$ **$x^2 + 6x + 9$**
$$(x + 3)^2 = x^2 + 2(x)(3) + 3^2$$
$$= x^2 + 6x + 9$$

10. $(2y - 1)^2 =$ **$4y^2 - 4y + 1$**
$$(2y - 1)^2 = (2y)^2 - 2(2y)(1) + 1^2$$
$$= 4y^2 - 4y + 1$$

Chapter 7: Using Special Rules for Multiplying Expressions

11 $(3a - 2b)^2 = 9a^2 - 12ab + 4b^2$

$$(3a - 2b)^2 = (3a)^2 - 2(3a)(2b) + (2b)^2$$
$$= 9a^2 - 12ab + 4b^2$$

12 $(5xy + z)^2 = 25x^2y^2 + 10xyz + z^2$

$$(5xy + z)^2 = (5xy)^2 + 2(5xy)(z) + z^2$$
$$= 25x^2y^2 + 10xyz + z^2$$

13 $(x + 3)(x - 3) = x^2 - 9$

14 $(2x - 7)(2x + 7) = 4x^2 - 49$

15 $(a^3 - 3)(a^3 + 3) = a^6 - 9$

$$(a^3 - 3)(a^3 + 3) = (a^3)^2 - 3^2 = a^6 - 9$$

16 $(2x^2h + 9)(2x^2h - 9) = 4x^4h^2 - 81$

$$(2x^2h + 9)(2x^2h - 9) = (2x^2h)^2 - 9^2 = 4x^4h^2 - 81$$

17 $(x + 1)^3 = x^3 + 3x^2 + 3x + 1$

$$(x + 1)^3 = x^3 + 3 \times x^2 \times 1 + 3 \times x \times 1^2 + 1^3$$
$$= x^3 + 3x^2 + 3x + 1$$

18 $(y - 2)^3 = y^3 - 6y^2 + 12y - 8$

$$(y - 2)^3 = y^3 - 3 \times y^2 \times 2 + 3 \times y \times 2^2 - 2^3 \text{ using } +, -, +, -$$
$$= y^3 - 6y^2 + 12y - 8$$

19 $(3z + 1)^3 = 27z^3 + 27z^2 + 9z + 1$

$$(3z + 1)^3 = (3z)^3 + 3(3z)^2 \times 1 + 3(3z) \times 1^2 + 1^3$$
$$= 27z^3 + 27z^2 + 9z + 1$$

20 $(5 - 2y)^3 = 125 - 150y + 60y^2 - 8y^3$

$$(5 - 2y)^3 = 5^3 - 3^1 \times 5^2 \times 2y + 3 \times 5 \times (2y)^2 - (2y)^3 \text{ using } +, -, +, -.$$
$$= 125 - 150y + 60y^2 - 8y^3$$

21 $(x - 2)(x^2 + 2x + 4) = x^3 - 8$

22 $(y + 1)(y^2 - y + 1) = y^3 + 1$

23 $(2z + 5)(4z^2 - 10z + 25) = 8z^3 + 125$

$$(2z + 5)(4z^2 - 10z + 25) = (2z)^3 + 5^3 \text{ by } (a + b)(a^2 - ab + b^2) = a^3 + b^3 \text{ with } a = 2z \text{ and } b = 5.$$
$$= 8z^3 + 125$$

24 $(3x - 2)(9x^2 + 6x + 4) = 27x^3 - 8$

$$(3x - 2)(9x^2 + 6x + 4) = (3x)^3 - 2^3 \text{ by } (a - b)(a^2 + ab + b^2) = a^3 - b^3.$$
$$= 27x^3 - 8$$

25 $(x+1)^4 = x^4 + 4x^3 + 6x^2 + 4x + 1$

$(x+1)^4 = x^4 + 4 \times x^3 \times 1 + 6 \times x^2 \times 1^2 + 4 \times x \times 1^3 + 1^4$
$= x^4 + 4x^3 + 6x^2 + 4x + 1$

26 $(2y-1)^4 = 16y^4 - 32y^3 + 24y^2 - 8y + 1$

$(2y-1)^4 = (2y+(-1))^4$
$= (2y)^4 + 4(2y)^3(-1) + 6(2y)^2(-1)^2 + 4(2y)(-1)^3 + (-1)^4$
$= 16y^4 - 32y^3 + 24y^2 - 8y + 1$

27 $(z-1)^5 = z^5 - 5z^4 + 10z^3 - 10z^2 + 5z - 1$

$(z-1)^5 = (z+(-1))^5$
$= z^5 + 5z^4(-1) + 10z^3(-1)^2 + 10z^2(-1)^3 + 5z(-1)^4 + (-1)^5$ (Use Pascal line 1-5-10-10-5-1.)
$= z^5 - 5z^4 + 10z^3 - 10z^2 + 5z - 1$

28 $(3z+2)^5 = 243z^5 + 810z^4 + 1080z^3 + 720z^2 + 240z + 32$

$(3z+2)^5 = (3z)^5 + 5(3z)^4 \times 2 + 10(3z)^3 \times 2^2 + 10(3z)^2 \times 2^3 + 5(3z) \times 2^4 + 2^5$
$= 243z^5 + 10(81z^4) + 40(27z^3) + 80(9z^2) + 80(3z) + 32$
$= 243z^5 + 810z^4 + 1080z^3 + 720z^2 + 240z + 32$

Chapter 8

Doing Long Division to Simplify Algebraic Expressions

In This Chapter
- Dividing by single terms
- Dividing by multiple terms
- Using synthetic division

Using long division to simplify expressions with numbers and variables has many similarities to using long division with just numbers. The variables do add an interesting twist (besides making everything look like alphabet soup) — with the exponents and different letters to consider. But you still have a *divisor, dividend,* and *quotient* (what divides in, what's divided into, and the answer). One major difference is that in algebra you usually write the remainders as fractions.

Dividing by a Monomial

Dividing an expression by a *monomial* (one term) can go one of two ways.

- You can divide every term in the expression by the divisor.
- You have one or more terms in the expression that don't divide evenly.

If a fraction can be divided — if every term can be divided by the divisor — it means that the denominator and numerator have the same common factor. For instance, in the first example in this section, the denominator, $6y$, divides every term in the numerator. First, factor out the numerator, and then reduce the fraction. You may want to have the expression divide evenly, but that isn't always the case.

Often, you'll have one or more terms in the expression — or the fraction's numerator — that won't contain all the factors in the divisor (denominator). In this case, writing the remainder as a fraction is fairly standard. Another option when dealing with fractions is to break up the problem into as many fractions as there are terms in the numerator. The method you use is pretty much dictated by what you want to do with the expression when you're done.

Part II: Operating and Factoring

EXAMPLE

Q. $\dfrac{24y^2 - 18y^3 + 30y^4}{6y} =$

A. $y(4 - 3y + 5y^2)$

$$\dfrac{24y^2 - 18y^3 + 30y^4}{6y} = \dfrac{6y^2(4 - 3y + 5y^2)}{6y}$$

$$= \dfrac{\cancel{6}y^{\cancel{2}1}(4 - 3y + 5y^2)}{\cancel{6}\cancel{y}}$$

$$= y(4 - 3y + 5y^2)$$

Q. $\dfrac{40x^4 - 32x^3 + 20x^2 - 12x + 3}{4x} =$

A. $10x^3 - 8x^2 + 5x - 3 + \tfrac{3}{4x}$

$$\dfrac{40x^4 - 32x^3 + 20x^2 - 12x + 3}{4x} = \dfrac{40x^4}{4x} - \dfrac{32x^3}{4x}$$
$$+ \dfrac{20x^2}{4x} - \dfrac{12x}{4x}$$
$$+ \dfrac{3}{4x}$$
$$= 10x^3 - 8x^2 + 5x - 3 + \dfrac{3}{4x}$$

1. $\dfrac{4x^3 - 3x^2 + 2x}{x} =$

Solve It

2. $\dfrac{8y^4 + 12y^5 - 16y^6 + 40y^8}{4y^4} =$

Solve It

3. $\dfrac{6x^5 - 2x^3 + 4x + 1}{x} =$

Solve It

4. $\dfrac{15x^3y^4 + 9x^2y^2 - 12xy}{3xy} =$

Solve It

Dividing by a Binomial

Dividing by a *binomial* (two terms) in algebra means that the two terms have to divide into another expression. When the division doesn't have a remainder, you know that the divisor was a factor of the original expression. When dividing a binomial into another expression, you always work toward getting rid of the *lead term* (the first term — the one with the highest power) in the binomial. See the following example for a clearer picture of this concept.

This example shows a dividend that starts with a third-degree term and then each of the terms in decreasing powers (second degree, first degree, and zero degree, which is a *constant* — just a number with no variable). If your dividend is missing any powers that are lower than the lead term, then you need to fill in the spaces with zeroes to keep your division lined up.

Also, if you have a remainder, remember to write that remainder as the numerator of a fraction with the divisor in the denominator.

Part II: Operating and Factoring

Q. $x-4 \overline{)x^3 - 9x^2 + 27x - 28} =$

A. $x^2 - 5x + 7$

1. **Get rid of the lead term, the x^3.**

   The value that you multiply the x by to get that lead term is x^2, so write it over the lead term in the quotient and then multiply each term in the divisor by that value:

   $$\begin{array}{r} x^2 \\ x-4 \overline{)x^3 - 9x^2 + 27x - 28} \\ x^3 - 4x^2 \end{array}$$

2. **Subtract the two values by multiplying (in algebra, *subtract* means to change the signs and add; see Chapter 2 for more information).**

   $$\begin{array}{r} x^2 \\ x-4 \overline{)x^3 - 9x^2 + 27x - 28} \\ -(x^3 - 4x^2) \\ \hline -5x^2 + 27x - 28 \end{array}$$

3. **Bring down the rest of the terms in the dividend, and you have a new lead term, $-5x^2$.**

4. **Multiply each term in the divisor by $-5x$ to match the new lead term.**

5. **Subtract again, and bring down the other terms.**

   $$\begin{array}{r} x^2 - 5x \\ x-4 \overline{)x^3 - 9x^2 + 27x - 28} \\ -(x^3 - 4x^2) \\ \hline -5x^2 + 27x - 28 \\ -(-5x^2 + 20x) \\ \hline 0 + 7x - 28 \end{array}$$

6. **Repeat until you run out of terms.**

   $$\begin{array}{r} x^2 - 5x + 7 \\ x-4 \overline{)x^3 - 9x^2 + 27x - 28} \\ -(x^3 - 4x^2) \\ \hline -5x^2 + 27x - 28 \\ +5x^2 - 20x \\ \hline 7x - 28 \\ 7x - 28 \\ \hline 0 \end{array}$$

   This problem divides evenly. It doesn't have a remainder, which means that $x-4$ is a factor of the dividend.

5. $x+2 \overline{) x^3 + 7x^2 + 3x - 14 } =$

Solve It

6. $x-3 \overline{) x^4 - 2x^3 - 5x^2 + 7x - 3 } =$

Solve It

7. $3x-4 \overline{) 12x^3 - 10x^2 - 17x + 12 } =$

Solve It

8. $x^2+1 \overline{) x^6 - 3x^5 + x^4 - 2x^3 - 3x^2 + x - 3 } =$

Solve It

Part II: Operating and Factoring

Dividing by Other Polynomials

Even though dividing by monomials or binomials is the most common situation in algebra, you may run across the occasional opportunity to divide by a *polynomial* with three or more terms. (Are you doing backflips and jumps of joy yet?)

Roll up your sleeves, grab your magic hat and wand, and get ready for some fun. The trick? Keep everything lined up correctly and fill in the blanks where they are missing powers.

Q. $(9x^6 - 4x^5 + 3x^2 - 1) \div (x^2 - 2x + 1) =$

A. $9x^4 + 14x^3 + 19x^2 + 24x + 32 + \dfrac{40x - 33}{x^2 - 2x + 1}$

Here's what the division looks like:

$$
\begin{array}{r}
9x^4 + 14x^3 + 19x^2 + 24x + 32 \\
x^2 - 2x + 1 \overline{\smash{)} 9x^6 - 4x^5 + 0 + 0 + 3x^2 + 0 - 1} \\
-(9x^6 - 18x^5 + 9x^4) \\
\hline
14x^5 - 9x^4 + 0 + 3x^2 + 0 - 1 \\
-(14x^5 - 28x^4 + 14x^3) \\
\hline
+19x^4 - 14x^3 + 3x^2 + 0 - 1 \\
-(19x^4 - 38x^3 + 19x^2) \\
\hline
+24x^3 - 16x^2 + 0 - 1 \\
-(24x^3 - 48x^2 + 24x) \\
\hline
+32x^2 - 24x - 1 \\
-(32x^2 - 64x + 32) \\
\hline
40x - 33
\end{array}
$$

9. $(x^4 - 2x^3 + x^2 - 7x - 2) \div (x^2 + 3x - 1) =$

Solve It

10. $(x^6 + 6x^4 - 4x^2 + 21) \div (x^4 - x^2 + 3) =$

Solve It

Trying Synthetic Division

Dividing by binomials is a very common procedure in algebra. The theory behind it allows you to determine factors of an expression and roots of an equation. A quick, easy way of dividing a polynomial by a binomial of the form $x + a$ or $x - a$ is called *synthetic division*. Notice that these two binomials have the variable to the first degree followed by a number being added or subtracted. Examples are $x + 4$ or $x - 7$.

To perform *synthetic division*, you use just the *coefficients* (the numbers multiplying the variables, x's) of the polynomial's terms being divided and the *opposite* of the number in the binomial.

Q. $(x^4 - 3x^3 + x - 4) \div (x + 1) =$

A. $x^3 - 4x^2 + 4x - 3 - \dfrac{1}{x+1}$

The opposite of the number in the divisor (–1 in this case) goes in front of the problem, in a little offset that looks like ⌐. Solve it this way:

1. **Write the coefficients in order, with a 0 to hold the place of the power 2 that's missing.**

2. **Bring down the first coefficient, below the horizontal line, and then multiply it times the number in front.**

3. **Write the result under the second coefficient, and add the two numbers; put the result on the bottom.**

4. **Take this new result, the –4 in this problem, and multiply it times the number in front, and then add the answer to the next coefficient.**

5. **Repeat this multiply-add process all the way down the line.**

The result on the bottom is the list of coefficients in the answer — plus the remainder, if you have one.

$$\begin{array}{r|rrrrr} -1 & 1 & -3 & 0 & 1 & -4 \\ & & -1 & 4 & -4 & 3 \\ \hline & 1 & -4 & 4 & -3 & -1 \end{array}$$

The answer uses the coefficients, in order, starting with one degree less than the polynomial that was divided. The last number is the remainder, and it goes over the divisor in a fraction.

11. $(x^4 - 2x^3 - 4x^2 + x + 6) \div (x - 3) =$

12. $(2x^4 + x^3 - 7x^2 + 5) \div (x + 2) =$

Answers to Problems on Division

This section provides the answers (in bold) to the practice problems in this chapter.

1. $\dfrac{4x^3 - 3x^2 + 2x}{x} = \mathbf{4x^2 - 3x + 2}$

$\dfrac{4x^3 - 3x^2 + 2x}{x} = \dfrac{\cancel{x}(4x^2 - 3x + 2)}{\cancel{x}} = 4x^2 - 3x + 2$

2. $\dfrac{8y^4 + 12y^5 - 16y^6 + 40y^8}{4y^4} = \mathbf{2 + 3y - 4y^2 + 10y^4}$

$\dfrac{8y^4 + 12y^5 - 16y^6 + 40y^8}{4y^4} = \dfrac{\cancel{4y^4}(2 + 3y - 4y^2 + 10y^4)}{\cancel{4y^4}}$

$= 2 + 3y - 4y^2 + 10y^4$

3. $\dfrac{6x^5 - 2x^3 + 4x + 1}{x} = \mathbf{6x^4 - 2x^2 + 4 + \dfrac{1}{x}}$

$\dfrac{6x^5 - 2x^3 + 4x + 1}{x} = \dfrac{6x^5}{x} - \dfrac{2x^3}{x} + \dfrac{4x}{x} + \dfrac{1}{x}$

$= 6x^4 - 2x^2 + 4 + \dfrac{1}{x}$

4. $\dfrac{15x^3y^4 + 9x^2y^2 - 12xy}{3xy} = \mathbf{5x^2y^3 + 3xy - 4}$

$\dfrac{15x^3y^4 + 9x^2y^2 - 12xy}{3xy} = \dfrac{\cancel{3xy}(5x^2y^3 + 3xy - 4)}{\cancel{3xy}}$

$= 5x^2y^3 + 3xy - 4$

5. $x+2 \,\overline{\smash{\big)}\, x^3 + 7x^2 + 3x - 14} = \mathbf{x^2 + 5x - 7}$

$$\begin{array}{r}
x^2 + 5x - 7 \\
x+2 \,\overline{\smash{\big)}\, x^3 + 7x^2 + 3x - 14} \\
\underline{-(x^3 + 2x^2)} \\
5x^2 + 3x \\
\underline{-(5x^2 + 10x)} \\
-7x - 14 \\
\underline{-(-7x - 14)} \\
0
\end{array}$$

6. $x-3 \,\overline{\smash{\big)}\, x^4 - 2x^3 - 5x^2 + 7x - 3} = \mathbf{x^3 + x^2 - 2x + 1}$

$$\begin{array}{r}
x^3 + x^2 - 2x + 1 \\
x-3 \,\overline{\smash{\big)}\, x^4 - 2x^3 - 5x^2 + 7x - 3} \\
\underline{-(x^4 - 3x^3)} \\
x^3 - 5x^2 \\
\underline{-(x^3 - 3x^2)} \\
-2x^2 + 7x \\
\underline{-(-2x^2 + 6x)} \\
x - 3 \\
\underline{-(x - 3)} \\
0
\end{array}$$

98 Part II: Operating and Factoring

7 $3x-4 \overline{\smash{)}12x^3-10x^2-17x+12} = \mathbf{4x^2+2x-3}$

$$\begin{array}{r} 4x^2+2x-3 \\ 3x-4 \overline{\smash{)}12x^3-10x^2-17x+12} \\ \underline{-(12x^3-16x^2)} \\ 6x^2-17x \\ \underline{-(6x^2-8x)} \\ -9x+12 \\ \underline{-(-9x+12)} \\ 0 \end{array}$$

8 $x^2+1 \overline{\smash{)}x^6-3x^5+x^4-2x^3-3x^2+x-3} = \mathbf{x^4-3x^3+x-3}$

$$\begin{array}{r} x^4-3x^3+x-3 \\ x^2+1 \overline{\smash{)}x^6-3x^5+x^4-2x^3-3x^2+x-3} \\ \underline{-(x^6+x^4)} \\ -3x^5-2x^3 \\ \underline{-(-3x^5-3x^3)} \\ x^3-3x^2+x \\ \underline{-(x^3+x)} \\ -3x^2-3 \\ \underline{-(-3x^2-3)} \\ 0 \end{array}$$

9 $(x^4-2x^3+x^2-7x-2) \div (x^2+3x-1) = \mathbf{x^2-5x+17} + \dfrac{-63x+15}{x^2+3x-1}$

$$\begin{array}{r} x^2-5x+17 \\ x^2+3x-1 \overline{\smash{)}x^4-2x^3+x^2-7x-2} \\ \underline{-(x^4+3x^3-x^2)} \\ -5x^3+2x^2-7x \\ \underline{-(-5x^3-15x^2+5x)} \\ 17x^2-12x-2 \\ \underline{-(17x^2+51x-17)} \\ -63x+15 \end{array}$$

10 $(x^6+6x^4-4x^2+21) \div (x^4-x^2+3) = \mathbf{x^2+7}$

$$\begin{array}{r} x^2+7 \\ x^4-x^2+3 \overline{\smash{)}x^6+0+6x^4+0-4x^2+0+21} \\ \underline{-(x^6-x^4+3x^2)} \\ 7x^4-7x^2+21 \\ \underline{-(7x^4-7x^2+21)} \\ 0 \end{array}$$

Write the remainder as a fraction with the divisor in the denominator of the fraction.

11 $(x^4-2x^3-4x^2+x+6) \div (x-3) = \mathbf{x^3+x^2-x-2}$

$$\begin{array}{r|rrrrr} \underline{3} & 1 & -2 & -4 & 1 & 6 \\ & & 3 & 3 & -3 & -6 \\ \hline & 1 & 1 & -1 & -2 & 0 \end{array}$$

So $x^3+x^2-x-2+\dfrac{0}{x-3}$

$=x^3+x^2-x-2$

Don't forget to use 0 as a placeholder when one of the powers is missing.

12 $(2x^4 + x^3 - 7x^2 + 5) \div (x + 2) =$ $2x^3 - 3x^2 - x + 2 + \dfrac{1}{x+2}$

Use the following breakdown to solve the problem:

$$\begin{array}{r|rrrrr} -2 & 2 & 1 & -7 & 0 & 5 \\ & & -4 & 6 & 2 & -4 \\ \hline & 2 & -3 & -1 & 2 & 1 \end{array}$$

Chapter 9
Factoring Algebraic Expressions

In This Chapter
- Writing prime factorizations of numbers
- Figuring out the greatest common factor
- Using factors to reduce algebraic fractions

*F*actoring an expression amounts to changing the form from addition and subtraction to multiplication and division. The change from an unfactored form to a factored form creates a single term — all tied together — so that you can perform other processes. Some of the other tasks in algebra that require a factored form are

- Reducing or simplifying fractions (see Chapter 3)
- Solving equations (see Chapters 12–15)
- Solving inequalities (see Chapter 16)
- Graphing functions (see Chapter 21)

Looking at Prime Factorizations

A *prime factorization* of a number is the product of all the prime numbers in that number. You first have to recognize which are the prime numbers. The first 25 prime numbers are 2, 3, 5, 7, 11, 13, 17, 19, 23, 29, 31, 37, 41, 43, 47, 53, 59, 61, 67, 71, 73, 79, 83, 89, and 97. Don't worry about memorizing them. Infinitely many more primes exist, of course, but these numbers are the most commonly used ones when doing prime factorizations.

To write the prime factorization of a number, start by writing that number as the product of two numbers and then writing the product of those two numbers and so on until you have only prime numbers in the product.

Q. Find the prime factorization of 360.

A. $360 = 2^3 \times 3^2 \times 5$

$360 = 10 \times 36 = 2 \times 5 \times 6 \times 6$
$= 2 \times 5 \times 2 \times 3 \times 2 \times 3$
$= 2^3 \times 3^2 \times 5$

Q. Find the prime factorization of 90.

A. $90 = 2 \times 3^2 \times 5$

$90 = 9 \times 10 = 3 \times 3 \times 2 \times 5 = 2 \times 3^2 \times 5$

You can start the multiplication in different ways. For example, maybe you started as the product of 6 and 15. How you start doesn't matter. You'll always end up with the same answer.

1. Write the prime factorization of 24.

Solve It

2. Write the prime factorization of 100.

Solve It

3. Write the prime factorization of 256.

Solve It

4. Write the prime factorization of 3,872.

Solve It

Factoring Out the Greatest Common Factor

The first line of attack when factoring an expression is to look for the *greatest common factor* (GCF), or the largest factor that will divide all terms evenly. You want a factor that divides each of the terms evenly — while, at the same time, not leaving any common factor in the resulting terms. You want to take out as many factors as possible.

Q. Find the GCF: $30x^4y^2 - 20x^5y^3 + 50x^6y$

A. $10x^4y$. If you divide each term by the greatest common factor and put the results of the divisions in a parenthesis, the factored result looks like this: $30x^4y^2 - 20x^5y^3 + 50x^6y = 10x^4y(3y - 2xy^2 + 5x^2)$

Q. Factor out the GCF: $8a^{3/2} - 12a^{1/2}$

A. $4a^{1/2}(2a - 3)$. Dealing with fractional exponents can be tricky. Just remember that the same rules apply to fractional exponents as with whole numbers. You subtract the exponents the same.

5. Factor out the GCF: $24x^2y^3 - 42x^3y^2$

Solve It

6. Factor out the GCF:
$11x^4(x-3) - 33x^2(x-3)^2 + 22x(x-3)^3$

Solve It

7. Factor out the GCF: $16a^2b^3c^4 - 48ab^4c^2$

Solve It

8. Factor out the GCF: $12y^{5/2} - 15y^{3/2}$

Solve It

Reducing Algebraic Fractions

The principles behind reducing fractions with numbers and reducing fractions with variables and numbers remain the same. You want to find something that divides both the *numerator* (the top of the fraction) and *denominator* (bottom of the fraction) evenly, and then leave the results of the division as the new numerator and denominator.

When the algebraic fraction has just multiplication and division in the numerator and denominator, the reducing part is pretty easy. Just divide out the common factors as shown in the following example.

Q. Reduce the fraction: $\dfrac{14x^3y}{21xy^4}$

A. $\dfrac{2x^2}{3y^3}$

When the fraction has two or more terms in the numerator, denominator, or both, you first have to factor out the GCF before you can reduce:

$$\dfrac{14x^3y}{21xy^4} = \dfrac{\overset{2}{\cancel{14}}x^{\cancel{3}2}\cancel{y}}{\underset{3}{\cancel{21}}\cancel{x}y^{\cancel{4}3}} = \dfrac{2x^2}{3y^3}$$

Q. Reduce the fraction: $\dfrac{15y^3 - 15y^2}{6y^5 - 6y^4}$

A. $\dfrac{5}{2y^2}$

$$\dfrac{15y^3 - 15y^2}{6y^5 - 6y^4} = \dfrac{15y^2(y-1)}{6y^4(y-1)} = \dfrac{\overset{5}{\cancel{15}}\cancel{y^2}\cancel{(y-1)}}{\underset{2}{\cancel{6}}y^{\cancel{4}2}\cancel{(y-1)}} = \dfrac{5}{2y^2}$$

9. Reduce the fraction: $\dfrac{33x^5}{66x^2}$

Solve It

10. Reduce the fraction: $\dfrac{15x^2y^3}{20xy^9}$

Solve It

11. Reduce the fraction: $\dfrac{14a^2b - 21a}{28ab^2}$

Solve It

12. Reduce the fraction:

$$\dfrac{6w^3(w+1)^1 - 8w^4(w+1)^3}{10w^5(w+1)^3}$$

Solve It

13. Reduce the fraction: $\dfrac{9009x^{4/3}y^2 - 7007x^{7/3}y}{4004x^{1/3}y}$

Solve It

14. Reduce the fraction:

$$\dfrac{8a^2b^3(c^2+1)^4 - 6a^3b^2(c^2+1)^3 + 14a^4b(c^2+1)^2}{4a^3b^2(c^2+1)^2 - 10a^4b^3(c^2+1)^3}$$

Solve It

Chapter 9: Factoring Algebraic Expressions

Answers to Problems on Factoring Expressions

This section provides the answers (in bold) to the practice problems in this chapter.

1. Write the prime factorization of 24. The answer is $2^3 \times 3$.

    $24 = 4 \times 6 = 2 \times 2 \times 2 \times 3 = 2^3 \times 3$

2. Write the prime factorization of 100. The answer is $2^2 \times 5^2$.

    $100 = 4 \times 25 = 2 \times 2 \times 5 \times 5 = 2^2 \times 5^2$

3. Write the prime factorization of 256. The answer is 2^8.

    $256 = 4 \times 64 = 4 \times 8 \times 8 = (2 \times 2)(2 \times 2 \times 2)(2 \times 2 \times 2) = 2^8$

4. Write the prime factorization of 3,872. The answer is $2^5 \times 11^2$.

    $3{,}872 = 4 \times 968 = 4 \times (8 \times 121) = (2 \times 2)(2 \times 2 \times 2)(11 \times 11) = 2^5 \times 11^2$

5. Factor out the GCF: $24x^2y^3 - 42x^3y^2$. The answer is $6x^2y^2(4y - 7x)$.

6. Factor out the GCF: $11x^4(x-3) - 33x^2(x-3)^2 + 22x(x-3)^3$.

    The answer is $11x(x-3)\,[x^3 - 3x\,(x-3) + 2\,(x-3)^2]$.

TIP

You may be able to simplify the expression in the brackets. You can distribute, square, and combine like terms. It depends on what you're planning on doing with this result — which determines how much you want to simplify those terms in the brackets.

7. Factor out the GCF: $16a^2b^3c^4 - 48ab^4c^2$. The answer is $16ab^3c^2\,(ac^2 - 3b)$.

8. Factor out the GCF: $12y^{5/2} - 15y^{3/2}$. The answer is $3y^{3/2}(4y - 5)$.

9. Reduce the fraction: $\dfrac{33x^5}{66x^2}$

    The answer is $\dfrac{x^3}{2}$.

10. Reduce the fraction: $\dfrac{15x^2y^3}{20xy^9}$

    The answer is $\dfrac{3x}{4y^6}$.

    $\dfrac{15x^2y^3}{20xy^9} = \dfrac{\cancel{15}^{\,3}x^2y^3}{\cancel{20}_{\,4}xy^9} = \dfrac{3x}{4y^6}$

11. Reduce the fraction: $\dfrac{14a^2b - 21a}{28ab^2}$

    The answer is $\dfrac{2ab - 3}{4b^2}$.

    $\dfrac{14a^2b - 21a}{28ab^2} = \dfrac{7a\,(2ab - 3)}{28ab^2} = \dfrac{2ab - 3}{4b^2}$

12. Reduce the fraction: $\dfrac{6w^3(w+1)^1 - 8w^4(w+1)^3}{10w^5(w+1)^3}$

    The answer is $\dfrac{3 - 4w\,(w+1)^2}{5w^2(w+1)^2}$.

    $\dfrac{6w^3(w+1)^1 - 8w^4(w+1)^3}{10w^5(w+1)^3} = \dfrac{2w^3(w+1)\bigl[3 - 4w\,(w+1)^2\bigr]}{10w^5(w+1)^3}$

    $= \dfrac{3 - 4w\,(w+1)^2}{5w^2(w+1)^2}$

106 Part II: Operating and Factoring

 Even though the answer appears to have a common factor in the numerator and denominator, you can't reduce it. The numerator has two terms, and the first term, the 3, doesn't have that common factor in it.

13 Reduce the fraction: $\dfrac{9009x^{4/3}y^2 - 7007x^{7/3}y}{4004x^{1/3}y}$. The answer is $\dfrac{x(9y-7x)}{4}$.

$$\dfrac{9009x^{4/3}y^2 - 7007x^{7/3}y}{4004x^{1/3}y} = \dfrac{1001x^{4/3}y(9y - 7x^1)}{4(1001)x^{1/3}y}$$

 When factoring the terms in the numerator, be careful with the subtraction of the fractions. These fractional exponents are found frequently in higher mathematics and behave just as well as you see here. I put the exponent of 1 on the x in the numerator just to emphasize the result of the subtraction of exponents. Continuing,

$$= \dfrac{\cancel{1001}x^{4/3\,1}\cancel{y}(9y - 7x^1)}{4\cancel{(1001)}x^{1/3}\cancel{y}} = \dfrac{x(9y-7x)}{4}$$

14 Reduce the fraction: $\dfrac{8a^2b^3(c^2+1)^4 - 6a^3b^2(c^2+1)^3 + 14a^4b(c^2+1)^2}{4a^3b^2(c^2+1)^2 - 10a^4b^3(c^2+1)^3}$. The answer is

$$\dfrac{4b^2(c^2+1)^2 - 3ab(c^2+1) + 7a^2}{ab\left[2 - 5ab(c^2+1)\right]}.$$

Factor the numerator and denominator separately, and then reduce by dividing by the common factors in each:

$$\dfrac{8a^2b^3(c^2+1)^4 - 6a^3b^2(c^2+1)^3 + 14a^4b(c^2+1)^2}{4a^3b^2(c^2+1)^2 - 10a^4b^3(c^2+1)^3}$$

$$= \dfrac{2a^2b(c^2+1)^2\left[4b^2(c^2+1)^2 - 3a^1b^1(c^2+1)^1 + 7a^2\right]}{2a^3b^2(c^2+1)^2\left[2 - 5ab(c^2+1)\right]}$$

$$= \dfrac{\cancel{2a^2b}\cancel{(c^2+1)^2}\left[4b^2(c^2+1)^2 - 3a^1b^1(c^2+1)^1 + 7a^2\right]}{\cancel{2}a^{\cancel{3}}b^{\cancel{2}}\cancel{(c^2+1)^2}\left[2 - 5ab(c^2+1)\right]}$$

$$= \dfrac{4b^2(c^2+1)^2 - 3ab(c^2+1) + 7a^2}{ab\left[2 - 5ab(c^2+1)\right]}$$

Chapter 10
Two at a Time with Factoring

In This Chapter
- Recognizing the difference between two squares
- Dealing with cubes — their sums and differences
- Using more than one factoring technique

You have several different ways to factor a *binomial* (an expression with two terms):

- You can factor out a greatest common factor (GCF).
- You can write the expression as the product of two binomials — one the sum of the two roots and the other the difference of those same two roots.
- You can factor an expression that has the sum or difference of two perfect cubes. You write the expression as the product of a binomial and *trinomial* (an expression with three terms) — one with the sum or difference of the cube roots and the other with squares of the roots and a product of the roots.

This chapter explains what you can do to change from two terms to one by factoring the expression. I cover GCF in Chapter 9. The other procedures are new to this chapter.

Factoring the Difference of Squares

When a binomial is the difference of two perfect squares, you can factor it into the product of the sum and difference of the square roots of those two terms.

$$a^2 - b^2 = (a+b)(a-b)$$

Q. Factor: $4x^2 - 81$

A. $(2x+9)(2x-9)$

Q. Factor: $25 - 36x^4y^2z^6$

A. $(5 + 6x^2yz^3)(5 - 6x^2yz^3)$. The square of $6x^2yz^3$ is $36x^4y^2z^6$.

Part II: Operating and Factoring

1. Factor: $x^2 - 25$

Solve It

2. Factor: $64a^2 - y^2$

Solve It

3. Factor: $49x^2y^2 - 9z^2w^4$

Solve It

4. Factor: $100x^{36} - 81y^{100}$

Solve It

Factoring Differences and Sums of Cubes

You can factor only the difference of two perfect squares. With cubes, though, both sums and differences factor into the product of a binomial and a trinomial.

$$a^3 - b^3 = (a-b)(a^2 + ab + b^2) \text{ and } a^3 + b^3 = (a+b)(a^2 - ab + b^2)$$

The pattern here is that first the sum or difference of cubes factors into the sum or difference of the two cube roots. Then that binomial is multiplied by a trinomial composed of the squares of those two cube roots and the *opposite* of the product of them. If the binomial has a + sign, the middle term of the trinomial is –. If the binomial has a – sign, then the middle term in the trinomial is +. The two squares in the trinomial are always positive.

Q. Factor: $x^3 - 27$

A. $(x-3)(x^2 + 3x + 9)$

$x^3 - 27 = (x-3)(x^2 + x \times 3 + 3^2) =$
$(x-3)(x^2 + 3x + 9)$

Q. Factor: $125 + 8y^3$

A. $(5 + 2y)(25 - 10y + 4y^2)$

$125 + 8y^3 = (5 + 2y)(5^2 - 5 \times 2y + [2y]^2) = (5 + 2y)(25 - 10y + 4y^2)$

5. Factor: $x^3 + 1$

Solve It

6. Factor: $8 - y^3$

Solve It

7. Factor: $27z^3 + 125$

Solve It

8. Factor: $8x^3 - 343y^6$

Solve It

Factoring in More Than One Way

Many factorization problems in mathematics involve multiple types of factoring. You may find a GCF in the terms, and then you may recognize that what's left is the difference of two cubes. You sometimes factor the difference of two squares just to find that one of those binomials is the difference of two new squares.

Solving these problems is really like figuring out a gigantic puzzle. You discover how to conquer it by applying the factorization rules. In general, first look for a GCF. Life is much easier when the numbers and powers are smaller — they're easier to deal with and work out in your head.

Q. Factor: $4x^6 + 108x^3$

A. $4x^3(x+3)(x^2 - 3x + 9)$

First, take out the GCF. Then factor the sum of the cubes in the parenthesis:

$4x^6 + 108x^3 =$
$4x^3(x^3 + 27) =$
$4x^3(x+3)(x^2 - 3x + 9)$

Q. Factor: $y^8 - 256$

A. $(y^4 + 16)(y^2 + 4)(y+2)(y-2)$

You can factor this problem as the difference of two squares. Then the second factor factors again and again:

$y^8 - 256 = (y^4 + 16)(y^4 - 16) =$
$(y^4 + 16)(y^2 + 4)(y^2 - 4) =$
$(y^4 + 16)(y^2 + 4)(y+2)(y-2)$

9. Completely factor: $3x^3y^3 - 27xy^3$

Solve It

10. Completely factor: $36x^2 - 100y^2$

Solve It

11. Completely factor: $80y^4 - 10y$

Solve It

12. Completely factor: $10,000x^4 - 1$

Solve It

13. Completely factor: $x^6 - 1$

Solve It

14. Completely factor: $125a^3b^3 - 125c^6$

Solve It

Answers to Problems on Factoring

This section provides the answers (in bold) to the practice problems in this chapter.

1. Factor: $x^2 - 25$. The answer is **$(x + 5)(x - 5)$**.

2. Factor: $64a^2 - y^2$. The answer is **$(8a + y)(8a - y)$**.

3. Factor: $49x^2y^2 - 9z^2w^4$. The answer is **$(7xy + 3zw^2)(7xy - 3zw^2)$**.

4. Factor: $100x^{36} - 81y^{100}$. The answer is **$(10x^{18} + 9y^{50})(10x^{18} - 9y^{50})$**.

5. Factor: $x^3 + 1$. The answer is **$(x + 1)(x^2 - x + 1)$**.

6. Factor: $8 - y^3$. The answer is **$(2 - y)(4 + 2y + y^2)$**.
   $8 - y^3 = (2 - y)(2^2 + 2y + y^2) = (2 - y)(4 + 2y + y^2)$

7. Factor: $27z^3 + 125$. The answer is **$(3z + 5)(9z^2 - 15z + 25)$**.
   $27z^3 + 125 = (3z + 5)\left((3z)^2 - 3z \times 5 + 5^2\right)$
   $= (3z + 5)(9z^2 - 15z + 25)$

8. Factor: $8x^3 - 343y^6$. The answer is **$(2x - 7y^2)(4x^2 + 14xy^2 + 49y^4)$**.
   Note that $7^3 = 343$.
   $8x^3 - 343y^6 = (2x - 7y^2)\left((2x)^2 + (2x)(7y^2) + (7y^2)^2\right)$
   $= (2x - 7y^2)(4x^2 + 14xy^2 + 49y^4)$

9. Completely factor: $3x^3y^3 - 27xy^3$. The answer is **$3xy^3(x + 3)(x - 3)$**.
   $3x^3y^3 - 27xy^3 = 3xy^3(x^2 - 9) = 3xy^3(x + 3)(x - 3)$

10. Completely factor: $36x^2 - 100y^2$. The answer is **$4(3x + 5y)(3x - 5y)$**.
    $36x^2 - 100y^2 = 4(9x^2 - 25y^2) = 4(3x + 5y)(3x - 5y)$

11. Completely factor: $80y^4 - 10y$. The answer is **$10y(2y - 1)(4y^2 + 2y + 1)$**.
    $80y^4 - 10y = 10y(8y^3 - 1)$
    $= 10y(2y - 1)\left((2y)^2 + 2y \times 1 + 1^2\right)$
    $= 10y(2y - 1)(4y^2 + 2y + 1)$

12. Completely factor: $10{,}000x^4 - 1$. The answer is **$(100x^2 + 1)(10x + 1)(10x - 1)$**.
    $10{,}000x^4 - 1 = (100x^2 + 1)(100x^2 - 1) = (100x^2 + 1)(10x + 1)(10x - 1)$

13. Completely factor: $x^6 - 1$. The answer is **$(x + 1)(x - 1)(x^4 + x^2 + 1)$**.
    $x^6 - 1 = (x^2 - 1)\left((x^2)^2 + x^2 \times 1 + 1^2\right)$
    $= (x^2 - 1)(x^4 + x^2 + 1)$
    $= (x + 1)(x - 1)(x^4 + x^2 + 1)$

    As a difference of squares, you could factor this problem as
    $x^6 - 1 = (x^3)^2 - 1 = (x^3 + 1)(x^3 - 1)$
    $= (x + 1)(x^2 - x + 1)(x + 1)(x^2 + x + 1)$

    Note that $x^4 + x^2 + 1 = (x^2 - x + 1)(x^2 + x + 1)$. This product isn't very obvious.

14. Completely factor: $125a^3b^3 - 125c^6$. The answer is $\mathbf{125(ab - c^2)(a^2b^2 + abc^2 + c^4)}$.

$$125a^3b^3 - 125c^6 = 125(a^3b^3 - c^6)$$
$$= 125(ab - c^2)\left((ab)^2 + (ab)c^2 + (c^2)^2\right)$$
$$= 125(ab - c^2)(a^2b^2 + abc^2 + c^4)$$

Chapter 11

Factoring Trinomials and Other Expressions

In This Chapter
▶ Locating the greatest common factor
▶ Factoring by "Un-FOIL-ing"
▶ Grouping terms to factor them
▶ Using multiple methods of factoring

In Chapter 10, you find the basic ways to factor a *binomial* (an expression with two terms). *Factoring* means to rewrite the expression all in one term, connected by multiplication and division. When dealing with a polynomial with four terms, such as $x^4 - 4x^3 - 11x^2 - 6x$, you write the factored form as $x(x + 1)^2(x - 6)$. The factored form has many advantages, especially when you want to simplify fractions, solve equations, or graph functions.

When working with an algebraic expression with three terms (a *trinomial*) or more terms, you have a number of different techniques to use when factoring it. You generally start with the greatest common factor (GCF) and then apply one or more of the other techniques. This chapter covers the different techniques and provides several sample questions for you to try.

Finding the Greatest Common Factor (GCF)

In any factoring problem, first you want to find a common factor. (You can look under your bed if you want, but I highly doubt it's there.) If you find one (in the problem), you then divide every term by that common factor and write the expression as the product of the common factor times all the results of the divisions.

The common factor can have more than one term.

Q. Factor out the GCF: $28x^2y - 21x^3y^2 + 35x^5y^3$

A. $7x^2y$. Dividing every term by that common factor, the resulting factored expression is $7x^2y(4 - 3xy + 5x^3y^2)$.

Q. Factor out the binomial GCF: $3(x - 5)^4 + 2a(x - 5)^3 - 11a^2(x - 5)^2$

A. Factoring out the square of the binomial, you get $(x - 5)^2[3(x - 5)^2 + 2a(x - 5) - 11a^2]$. This example has a varying number of factors, or powers, of the binomial $(x - 5)$ in each term.

Part II: Operating and Factoring

1. Factor out the GCF: $8x^3y^2 - 4x^2y^3 + 14xy^4$

Solve It

2. Factor out the GCF: $36w^4 - 24w^3 - 48w^2$

Solve It

3. Factor out the GCF:
$15(x-3)^3 + 60x^4(x-3)^2 + 5(x-3)$

Solve It

4. Factor out the GCF:
$5abcd + 10a^2bcd + 30bcde$

Solve It

Chapter 11: Factoring Trinomials and Other Expressions

"Un"wrapping the FOIL

You use the FOIL method to help when multiplying two binomials. (See Chapter 7 for some problems that use this method.) You use the *unFOIL* method, which re-creates the two binomials that were multiplied together, when you're faced with a trinomial with a squared term in it. The task in this chapter is to *factor* trinomials that have been created by multiplying two binomials together — figure out what those binomials are and write them as a product.

The general procedure for performing unFOIL includes these steps:

1. **Write the trinomial in descending powers of a variable.**
2. **Find all the combinations of factors whose product is the first term in the trinomial.**
3. **Find all the sets of factors whose product is the last term in the trinomial.**
4. **Try different combinations of those factors in the binomials so that the middle term is the result of combining the outer and inner products.**

Q. Factor $2x^2 - 5x - 3$.

A. $(2x + 1)(x - 3)$

1. Factor the first term into $2x$ times x.
2. Factor the last term into 3 times 1.
3. Decide on the middle term — sum or difference.

   Because the last term is *negative*, you want to find a way to arrange the factors so that the outer and inner products have the sum of $-5x$. Arrange them this way: $(2x$ and $1)(x$ and $3)$.

4. Decide on the placement of the signs.

   To enter the signs, you need a $-$ and $+$, because the last term is negative and you need two different signs. Putting the $-$ sign in front of the 3 results in a $-6x$ and a $+1x$. Combining them gives you the $-5x$.

Q. Factor $12y^2 - 17y + 6$.

A. $(4y - 3)(3y - 2)$. The factors of the first term are either y and $12y$, $2y$ and $6y$, or $3y$ and $4y$. The factors of the last term are either 1 and 6 or 2 and 3. The last term is positive, so the outer and inner products have to have a sum of $17y$.

5. Factor $x^2 - 8x + 15$.

Solve It

6. Factor $y^2 - 6y - 40$.

Solve It

7. Factor $2x^2 + 3x - 2$.

Solve It

8. Factor $4z^2 + 12z + 9$.

Solve It

Chapter 11: Factoring Trinomials and Other Expressions

9. Factor $w^2 - 16$.

Solve It

10. Factor $12x^2 - 8x - 15$.

Solve It

11. Factor $x^{10} + 4x^5 + 3$.

Solve It

12. Factor $4y^{16} - 9$.

Solve It

Factoring Trinomials in More Than One Way

Trinomials can be factored by taking out a GCF or by using the *un-FOIL* method. Sometimes you can use both of these methods in one expression. When this happens, first take out the common factor to make the terms in the expression simpler and the numbers smaller.

Q. Factor $9x^4 - 18x^3 - 72x^2$.

A. $9x^2(x - 4)(x + 2)$. First take out the common factor and write the product with that common factor outside the parenthesis: $9x^2(x^2 - 2x - 8)$. You can then factor the trinomial inside the parenthesis.

Q. Factor $(x - 3)^3 + (x - 3)^2 - 30(x - 3)$.

A. $(x - 3)[(x - 8)(x + 3)]$. Remember, the common factor can be a binomial. First, factor out the binomial $(x - 3)$, which gives you $(x - 3)[(x - 3)^2 + (x - 3) - 30]$. Square the first term in the brackets, and then simplify the expression in the brackets by combining like terms: $(x - 3)[x^2 - 6x + 9 + x - 3 - 30] = (x - 3)[x^2 - 5x - 24]$. Now you can factor the trinomial in the brackets.

13. Completely factor: $3z^2 - 12z + 12$

Solve It

14. Completely factor: $5y^3 - 5y^2 - 10y$

Solve It

15. Completely factor: $x^6 - 18x^5 + 81x^4$

Solve It

16. Completely factor: $w^4 - 10w^2 + 9$

Solve It

17. Completely factor:
$3x^2(x-2)^2 + 9x(x-2)^2 - 12(x-2)^2$

Solve It

18. Completely factor:
$a^2(x^2-25) - 15a(x^2-25) + 14(x^2-25)$

Solve It

Factoring by Grouping

The expression $2axy + 8x - 3ay - 12$ has four terms — terms that don't share a single common factor. The first two terms have a common factor of $2x$, and the last two terms have a common factor of -3. What to do, what to do!

Don't worry. This problem suggests that factoring by *grouping* may be an option. In order to take this step, though, the grouped factoring has to result in a common factor in the individual groups. Check out the following example.

Q. Factor by grouping: $2axy + 8x - 3ay - 12$

A. $(ay + 4)(2x - 3)$. Factor $2x$ out of the first two terms and -3 out of the second two terms: $2x(ay + 4) - 3(ay + 4)$. You can see how you now have two terms, instead of four, and the two terms have a common factor of $(ay + 4)$. If you factor out a 3 instead of a -3, the second term would have been $(-ay - 4)$ for the second factor, which isn't the same as $(ay + 4)$. For factoring by grouping to work, the two new common factors have to be exactly the same.

Q. Factor by grouping: $2a^3x - a^3y - 6b^2x + 3b^2y + 2cx - cy$

A. $(2x - y)(a^3 - 3b^2 + c)$. Remember, this type of factoring by grouping can also work with six terms. First, find the common factor in each pair of terms. Factor a^3 out of the first two terms, $-3b^2$ out of the second two terms, and c out of the last two: $a^3(2x - y) - 3b^2(2x - y) + c(2x - y)$. Notice that, with the middle pairing, if you factor out $3b^2$ instead of $-3b^2$, you don't have the same common factor as the other two pairings — the signs are wrong.

19. Factor by grouping: $ab^2 + 2ab + b + 2$

Solve It

20. Factor by grouping: $xz^2 - 5z^2 + 3x - 15$

Solve It

Chapter 11: Factoring Trinomials and Other Expressions

21. Factor by grouping: $mn - 3m - 4n + 12$

Solve It

22. Factor by grouping: $abcd - 4ab + 2cd - 8$

Solve It

23. Factor by grouping:
$ax - 3x + ay - 3y + az - 3z$

Solve It

24. Factor by grouping:
$x^2y^2 + 3y^2 + x^2y + 3y - 6x^2 - 18$

Solve It

Putting All the Factoring Together

Factoring an algebraic expression is somewhat like styling hair. You categorize the job first depending on how many terms you have (the same as looking at length and texture of the tresses) — that amount tells you what options you have for factoring (or the applications needed). If you have two terms, then you might get to factor them as the difference of squares or cubes or the sum of cubes. If you have three terms, then you may be able to factor the expression as the product of two binomials. With four or more terms, grouping may work. In any case, you first want to look for the GCF. Whether factoring or styling hair, you want an attractive result when you're finished — with minimum frizz.

Q. Completely factor: $8x^3 + 56x^2 - 240x$

A. $x^2(x + 10)(x - 3)$. The example problem has a common factor of $8x$. First, factor that out: $8x(x^2 + 7x - 30)$. Then you can use the un-FOIL method on the trinomial in the parenthesis. Finally, write the answer as the factor $8x$ times the product of the two binomials.

Q. Completely factor: $3x^5 - 75x^3 + 24x^2 - 600$

A. $3[(x + 5)(x - 5)(x + 2)(x^2 - 2x + 4)]$

1. **Factor out the common factor of 3.**

   $3(x^5 - 25x^3 + 8x^2 - 200)$

2. **Factor by grouping.**

   $3[x^3(x^2 - 25) + 8(x^2 - 25)] =$
   $3[(x^2 - 25)(x^3 + 8)]$

   The first factor in the brackets is the difference of perfect squares. The second factor is the sum of perfect cubes.

3. **Factor each of the two binomials in the parenthesis to find the answer.**

   Refer to Chapter 10 if you need a refresher on those factorizations.

25. Completely factor: $5x^3 - 80x$

Solve It

26. Completely factor: $y^5 - 9y^3 + y^2 - 9$

Solve It

27. Completely factor: $3x^5 - 66x^3 - 225x$

Solve It

28. Completely factor: $z^6 - 64$

Solve It

29. Completely factor: $8a^3b^2 - 32a^3 - b^2 + 4$

Solve It

30. Completely factor: $z^8 - 97z^4 + 1{,}296$. **Hint:** $1{,}296 = 81 \times 16$

Solve It

31. Completely factor: $4m^5 - 4m^4 - 36m^3 + 36m^2$

Solve It

32. Completely factor: $10y^7 + 350y^4 + 2{,}160y$

Solve It

Answers to Problems on Factoring Trinomials and Other Expressions

This section provides the answers (in bold) to the practice problems in this chapter.

1. Factor out the GCF: $8x^3y^2 - 4x^2y^3 + 14xy^4$. The answer is $\mathbf{2xy^2(4x^2 - 2xy + 7y^2)}$.

2. Factor out the GCF: $36w^4 - 24w^3 - 48w^2$. The answer is $\mathbf{12w^2(3w^2 - 2w - 4)}$.

3. Factor out the GCF: $15(x-3)^3 + 60x^4(x-3)^2 + 5(x-3)$.
   The answer is $\mathbf{5(x-3)[12x^5 - 36x^4 + 3x^2 - 18x + 28]}$.
   $$15(x-3)^3 + 60x^4(x-3)^2 + 5(x-3) = 5(x-3)\left[3(x-3)^2 + 12x^4(x-3) + 1\right]$$
   $$= 5(x-3)\left[12x^5 - 36x^4 + 3x^2 - 18x + 28\right]$$

4. Factor out the GCF: $5abcd + 10a^2bcd + 30bcde$. The answer is $\mathbf{5bcd(a + 2a^2 + 6e)}$.

5. Factor $x^2 - 8x + 15$. The answer is $\mathbf{(x-5)(x-3)}$, considering -15 and -1, or -5 and -3 for the factors of 15.

6. Factor $y^2 - 6y - 40$. The answer is $\mathbf{(y-10)(y+4)}$, which needs opposite signs with factors of 40 to be either 40 and 1, 20 and 2, 10 and 4, or 8 and 5.

7. Factor $2x^2 + 3x - 2$. The answer is $\mathbf{(2x-1)(x+2)}$.

8. Factor $4z^2 + 12z + 9$. The answer is $\mathbf{(2z+3)^2}$.
   $4z^2 + 12z + 9 = (2z+3)(2z+3) = (2z+3)^2$

9. Factor $w^2 - 16$. The answer is $\mathbf{(w+4)(w-4)}$. This is a difference of squares.

10. Factor $12x^2 - 8x - 15$. The answer is $\mathbf{(6x+5)(2x-3)}$.

11. Factor $x^{10} + 4x^5 + 3$. The answer is $\mathbf{(x^5+3)(x^5+1)}$.

12. Factor $4y^{16} - 9$. The answer is $\mathbf{(2y^8+3)(2y^8-3)}$. This is a difference of squares.

13. Completely factor: $3z^2 - 12z + 12$. The answer is $\mathbf{3(z-2)^2}$.
    $3z^2 - 12z + 12 = 3(z^2 - 4z + 4) = 3(z-2)(z-2) = 3(z-2)^2$

14. Completely factor: $5y^3 - 5y^2 - 10y$. The answer is $\mathbf{5y(y-2)(y+1)}$.
    $5y^3 - 5y^2 - 10y = 5y(y^2 - y - 2) = 5y(y-2)(y+1)$

15. Completely factor: $x^6 - 18x^5 + 81x^4$. The answer is $\mathbf{x^4(x-9)^2}$.
    $x^6 - 18x^5 + 81x^4 = x^4(x^2 - 18x + 81) = x^4(x-9)(x-9) = x^4(x-9)^2$

16. Completely factor: $w^4 - 10w^2 + 9$. The answer is $\mathbf{(w+3)(w-3)(w+1)(w-1)}$.
    $w^4 - 10w^2 + 9 = (w^2-9)(w^2-1) = (w+3)(w-3)(w+1)(w-1)$

17. Completely factor: $3x^2(x-2)^2 + 9x(x-2)^2 - 12(x-2)^2$. The answer is $\mathbf{3(x-2)^2(x+4)(x-1)}$.
    $$3x^2(x-2)^2 + 9x(x-2)^2 - 12(x-2)^2 = 3(x-2)^2\left[x^2 + 3x - 4\right]$$
    $$= 3(x-2)^2(x+4)(x-1)$$

18 Completely factor: $a^2(x^2 - 25) - 15a(x^2 - 25) + 14(x^2 - 25)$.

The answer is **$(x + 5)(x - 5)(a - 14)(a - 1)$**.

$$a^2(x^2 - 25) - 15a(x^2 - 25) + 14(x^2 - 25) = (x^2 - 25)[a^2 - 15a + 14]$$
$$= (x + 5)(x - 5)(a - 14)(a - 1)$$

19 Factor by grouping: $ab^2 + 2ab + b + 2$. The answer is **$(b + 2)(ab + 1)$**.

$ab^2 + 2ab + b + 2 = ab(b + 2) + 1(b + 2) = (b + 2)(ab + 1)$

20 Factor by grouping: $xz^2 - 5z^2 + 3x - 15$. The answer is **$(x - 5)(z^2 + 3)$**.

$xz^2 - 5z^2 + 3x - 15 = z^2(x - 5) + 3(x - 5) = (x - 5)(z^2 + 3)$

21 Factor by grouping: $mn - 3m - 4n + 12$. The answer is **$(n - 3)(m - 4)$**.

$mn - 3m - 4n + 12 = m(n - 3) - 4(n - 3) = (n - 3)(m - 4)$

22 Factor by grouping: $abcd - 4ab + 2cd - 8$. The answer is **$(cd - 4)(ab + 2)$**.

$abcd - 4ab + 2cd - 8 = ab(cd - 4) + 2(cd - 4) = (cd - 4)(ab + 2)$

23 Factor by grouping: $ax - 3x + ay - 3y + az - 3z$. The answer is **$(a - 3)(x + y + z)$**.

$$ax - 3x + ay - 3y + az - 3z = x(a - 3) + y(a - 3) + z(a - 3)$$
$$= (a - 3)(x + y + z)$$

24 Factor by grouping: $x^2y^2 + 3y^2 + x^2y + 3y - 6x^2 - 18$. The answer is **$(x^2 + 3)(y + 3)(y - 2)$**.

$$x^2y^2 + 3y^2 + x^2y + 3y - 6x^2 - 18 = y^2(x^2 + 3) + y(x^2 + 3) - 6(x^2 + 3)$$
$$= (x^2 + 3)(y^2 + y - 6)$$
$$= (x^2 + 3)(y + 3)(y - 2)$$

25 Completely factor: $5x^3 - 80x$. The answer is **$5x(x + 4)(x - 4)$**. First factor $5x$ out of each term, then you can factor the binomial as the sum and difference of the same two terms.

26 Completely factor: $y^5 - 9y^3 + y^2 - 9$. The answer is **$(y + 3)(y - 3)(y + 1)(y^2 - y + 1)$**.

$$y^5 - 9y^3 + y^2 - 9 = y^3(y^2 - 9) + 1(y^2 - 9)$$
$$= (y^2 - 9)(y^3 + 1)$$
$$= (y + 3)(y - 3)(y + 1)(y^2 - y + 1)$$

27 Completely factor: $3x^5 - 66x^3 - 225x$. The answer is **$3x(x + 5)(x - 5)(x^2 + 3)$**.

$$3x^5 - 66x^3 - 225x = 3x(x^4 - 22x^2 - 75)$$
$$= 3x(x^2 - 25)(x^2 + 3)$$
$$= 3x(x + 5)(x - 5)(x^2 + 3)$$

28 Completely factor: $z^6 - 64$. The answer is **$(z + 2)(z^2 - 2z + 4)(z - 2)(z^2 + 2z + 4)$**.

$z^6 - 64 = (z^3 + 8)(z^3 - 8) = (z + 2)(z^2 - 2z + 4)(z - 2)(z^2 + 2z + 4)$

Do this problem as the difference of cubes rather than the difference of squares. The complete factoring is easier.

Chapter 11: Factoring Trinomials and Other Expressions

29 Completely factor: $8a^3b^2 - 32a^3 - b^2 + 4$. The answer is $(b + 2)(b - 2)(2a - 1)(4a^2 + 2a + 1)$.

$$\begin{aligned}8a^3b^2 - 32a^3 - b^2 + 4 &= 8a^3(b^2 - 4) - 1(b^2 - 4) \\ &= (b^2 - 4)(8a^3 - 1) \\ &= (b + 2)(b - 2)(2a - 1)(4a^2 + 2a + 1)\end{aligned}$$

30 Completely factor: $z^8 - 97z^4 + 1{,}296$. The answer is $(z^2 + 9)(z + 3)(z - 3)(z^2 + 4)(z + 2)(z - 2)$.

$$\begin{aligned}z^8 - 97z^4 + 1{,}296 &= (z^4 - 81)(z^4 - 16) \\ &= (z^2 + 9)(z^2 - 9)(z^2 + 4)(z^2 - 4) \\ &= (z^2 + 9)(z + 3)(z - 3)(z^2 + 4)(z + 2)(z - 2)\end{aligned}$$

31 Completely factor: $4m^5 - 4m^4 - 36m^3 + 36m^2$. The answer is $4m^2(m - 1)(m + 3)(m - 3)$.

$$\begin{aligned}4m^5 - 4m^4 - 36m^3 + 36m^2 &= 4m^2(m^3 - m^2 - 9m + 9) \\ &= 4m^2\left[m^2(m - 1) - 9(m - 1)\right] \\ &= 4m^2\left[(m - 1)(m^2 - 9)\right] \\ &= 4m^2(m - 1)(m + 3)(m - 3)\end{aligned}$$

32 Completely factor: $10y^7 + 350y^4 + 2{,}160y$. The answer is $10y(y + 2)(y^2 - 2y + 4)(y + 3)(y^2 - 3y + 9)$.

$$\begin{aligned}10y^7 + 350y^4 + 2{,}160y &= 10y\left(y^6 + 35y^3 + 216\right) \\ &= 10y\left(y^6 + 35y^3 + 8 \times 27\right) \\ &= 10y\left(y^3 + 8\right)\left(y^3 + 27\right) \\ &= 10y(y + 2)\left(y^2 - 2y + 4\right)(y + 3)\left(y^2 - 3y + 9\right)\end{aligned}$$

Part III
Stirring Up Solutions

"He ran some linear equations, threw in a few theorems, and before I knew it I was buying rustproofing."

In This Part . . .

Everyone likes to be the one to solve the prickly problem that has stumped others. And you don't have to be a Sherlock Holmes. Finding the solution or solutions of an equation or inequality takes the three P's: patience, perseverance, and a plan. Remember that you need all three. You can be as patient as you want, but a solution isn't going to come to you if you aren't willing to diligently carry on with a thought-out plan. In these chapters you see how to recognize the different types of problems and decide on the best approaches. You then can apply the different algebraic techniques— and see why they are so important.

Chapter 12

Putting It on the Line: Solving Linear Equations

In This Chapter
- Solving equations using basic operations
- Working with grouping symbols and fractions
- Dealing with proportions

*L*inear equations are algebraic equations that have one or more variables and no powers greater than the first power. The most common way of solving linear equations is to perform operations or other manipulations so that the variable you're solving for is on one side of the equation and the numbers or other letters and symbols are on the other side of the equation. You want to get the variable alone so that you can finish with a statement, such as $x = 4$ or $y = 2a$. Because linear equations involve just the first degree (power one) of the variable, you'll get just one answer.

This chapter provides you tons of questions to hone your linear-equation skills. Come on in. The water is warm.

Using the Addition/Subtraction Property

One of the properties of equations is that you can add or subtract the same number from each side and not change the equation's *equality*. The equation is still a true statement (as long as it started out that way). You add the same number to each side (adding a negative is like subtracting the same thing from each side). You use this property to get the variables all to one side and the numbers and all the other letters and numbers to the other side.

You can check the solution by putting the answer back in the original equation to see if it works.

Q. Solve for x: $x + 7 = 11$

A. $x = 4$. Add -7 to each side (subtract 7 from each side). It looks like this:

$$\begin{array}{r} x + 7 = 11 \\ -7 -7 \\ \hline x = 4 \end{array}$$

Part III: Stirring Up Solutions

Q. Solve for y: $8y - 2 = 7y - 10$

A. $y = -8$. First add $-7y$ to each side to get rid of the variable on the right, and then you add 2 to each side to get the numbers on the right. This is what the process looks like:

$$\begin{aligned} 8y - 2 &= 7y - 10 \\ -7y & -7y \\ \hline y - 2 &= -10 \\ +2 & +2 \\ \hline y &= -8 \end{aligned}$$

1. Solve for x: $x + 4 = 15$

Solve It

2. Solve for y: $y - 2 = 11$

Solve It

3. Solve for x: $5x + 3 = 4x - 1$

Solve It

4. Solve for y: $2y + 9 + 6y - 8 = 4y + 5 + 3y - 11$

Solve It

Using the Multiplication/Division Property

Another property of equations is that when you multiply or divide both sides by the same number (not 0), then the equation is still an *equality* — the equation is still true. You can use this property to help solve equations for the value of the variable.

Q. Solve for x: $-3x = -45$

A. $x = 15$. Divide each side by -3 to determine what x is:
$$\frac{-3x}{-3} = \frac{-45}{-3}$$
$$x = 15$$

Q. Solve for y: $\frac{y}{5} = 12$

A. $y = 60$. Multiply each side by 5 to solve the equation for y:
$$\frac{y}{5} = 12$$
$$5 \times \frac{y}{5} = 12 \times 5$$
$$y = 60$$

5. Solve for x: $6x = 24$

Solve It

6. Solve for y: $-4y = 20$

Solve It

7. Solve for z: $\frac{z}{3} = 11$

Solve It

8. Solve for w: $\frac{w}{-4} = -2$

Solve It

Putting Several Operations Together

The different properties of equations that allow you to add the same number to each side or multiply each side by the same number (except 0) are the backbone of solving linear equations. More often than not, you have to perform several different operations to solve a particular variable. When the problem just has the operations of addition, subtraction, multiplication, and division, and doesn't have any grouping symbols to change the rules, you first do all the addition and subtraction to get the variables on one side and the numbers on the other side. Then you can multiply or divide to get the variable by itself.

The side you move the variable to really doesn't matter. Many people like to have it on the left, so you can read $x = 2$ as "x equals 2". Writing $2 = x$ is just as correct. You may prefer having the variable on the left if it makes for less awkward work or keeps the variable with a positive factor.

Q. Solve for x: $8x - 3 = 5x + 9$

A. $x = 4$

1. Add 3 and subtract $5x$ from each side.
2. Divide each side by 3 to get $x = 4$.

$$\begin{array}{r} 8x - 3 = 5x + 9 \\ -5x + 3 \quad -5x + 3 \\ \hline 3x = 12 \\ \frac{3x}{3} = \frac{12}{3} \\ x = 4 \end{array}$$

Q. Solve for y: $\frac{2y}{3} + 1 = \frac{4y}{3} + 5$

A. $-6 = y$

1. By subtracting $\frac{2y}{3}$ from each side and subtracting 5 from each side, you get

$$\begin{array}{r} \frac{2y}{3} + 1 = \frac{4y}{3} + 5 \\ -\frac{2y}{3} - 5 \quad -\frac{2y}{3} - 5 \\ \hline -4 = \frac{2y}{3} \end{array}$$

2. Multiply each side by 3.
3. Divide each side by 2.

Another way to do the last two steps in just one is to multiply by the reciprocal of ⅔, which is 3/2:

$$3(-4) = \frac{2y}{3} \times 3$$
$$-12 = 2y$$
$$\frac{-12}{2} = \frac{2y}{2}$$
$$-6 = y$$

9. Solve for x: $3x - 4 = 5$

Solve It

10. Solve for y: $8 - \frac{y}{2} = 7$

Solve It

11. Solve for x: $5x - 3 = 8x + 9$

Solve It

12. Solve for z: $\frac{z}{6} - 3 = z + 7$

Solve It

13. Solve for y: $4y + 16 - 3y = 7 + 3y$

Solve It

14. Solve for x: $\frac{3x}{4} - 2 = \frac{9x}{4} + 13$

Solve It

Solving Linear Equations with Grouping Symbols

In general, when solving linear equations, you add and subtract first and then multiply or divide. This general rule is interrupted when the problem has grouping symbols such as ( ), [ ], { }. See Chapter 2 for more on grouping symbols. If the equation has grouping symbols, you need to perform whatever operation is indicated by the grouping symbol before carrying on with the other rules. Also, if you perform an operation on the grouping symbol, then every term in the grouping symbol has to have that operation performed on it.

Fractions can act like grouping symbols, too. If the denominator doesn't divide all the terms in the numerator evenly, then get rid of the denominator by multiplying that fraction and all the terms in the equation by the denominator.

Q. Solve $8(2x + 1) + 6 = 5(x - 3) + 7$.

A. $x = -2$

1. Distribute the 8 over the two terms in the first parenthesis and the 5 over the two terms in the second parenthesis.

    You then get the equation $16x + 8 + 6 = 5x - 15 + 7$.

2. Combine the two numbers on each side to get the equation $16x + 14 = 5x - 8$.

3. Then subtract $5x$ from each side, subtract 14 from each side, and divide each side by 11.

$$16x + 14 = 5x - 8$$
$$-5x - 14 \quad -5x - 14$$
$$\overline{11x \quad = \quad -22}$$
$$\frac{11x}{11} = \frac{-22}{11}$$
$$x = -2$$

Q. Solve $\frac{x-5}{4} + 3 = x + 4$.

A. $x = -3$. First multiply each term on both sides of the equation by 4:

$$\left(4 \times \frac{x-5}{4}\right) + (4 \times 3) = (4 \times x) + (4 \times 4)$$
$$x - 5 + 12 = 4x + 16$$
$$x + 7 = 4x + 16$$

Now subtract $4x$ from each side and subtract 7 from each side and divide each side by -3.

$$x + 7 = 4x + 16$$
$$-4x - 7 \quad -4x - 7$$
$$\overline{-3x \quad = \quad 9}$$
$$\frac{-3x}{-3} = \frac{9}{-3}$$
$$x = -3$$

15. Solve for x: $3(x-5) = 12$

Solve It

16. Solve for y: $4(y+3) + 7 = 3$

Solve It

17. Solve for x: $\dfrac{4x+1}{3} = x+2$

Solve It

18. Solve for y: $5(y-3) - 3(y+4) = 1 - 6(y-4)$

Solve It

19. Solve for x: $x(3x + 1) - 2 = 3x^2 - 5$

Solve It

20. Solve for x: $(x - 3)(x + 4) = (x + 1)(x - 2)$

Solve It

Working It Out with Fractions

Fractions aren't everyone's favorite thing — although you wouldn't be able to get along without them in life or in algebra. Fractions in algebraic equations can complicate everything, so getting rid of them is often easier than trying to deal with finding common denominators several times in the same problem.

You need to consider two general procedures when working out fractions.

- If you can easily isolate the term with the fraction on one side, do the necessary addition and subtraction, and then multiply each side of the equation by the denominator of the fraction (be sure to multiply *each* term by that denominator).

- If the equation has more than one fraction, then find a common denominator for *all* the terms, and then multiply each side by this common denominator — in effect, getting rid of all the fractions.

Q. Use the first procedure from the preceding bulleted list to solve $\frac{2x-1}{3} + 4 = 7$.

A. $x = 5$

1. Subtract 4 from each side.
   $$\frac{2x-1}{3} = 3$$

2. Multiply each side by 3.
   $$3 \times \frac{2x-1}{3} = 3 \times 3$$
   $$2x - 1 = 9$$

3. Now it's in a form ready to solve by adding 1 to each side and dividing each side by 2.
   $$2x - 1 = 9$$
   $$+1 +1$$
   $$\overline{2x = 10}$$
   $$\frac{2x}{2} = \frac{10}{2}$$
   $$x = 5$$

Q. Now, use the second procedure from the preceding bulleted list to solve $\frac{x}{4} + 7 = 1 - \frac{x}{2}$.

A. $x = -8$

1. Determine the common denominator for the fractions, which is 4.

2. Multiply each term in the equation by 4.
   $$\left(4 \times \frac{x}{4}\right) + (4 \times 7) = (4 \times 1) - \left(4 \times \frac{x}{2}\right)$$
   which eliminates all the fractions, when the fractions are reduced.
   $$\left(\cancel{4} \times \frac{x}{\cancel{4}}\right) + (4 \times 7) = (4 \times 1) - \left(\overset{2}{\cancel{4}} \times \frac{x}{\cancel{2}}\right)$$
   $$x + 28 = 4 - 2x$$

3. Add $2x$ to each side and subtract 28 from each side.

4. Finish solving the equation.
   $$\begin{aligned} x + 28 &= 4 - 2x \\ +2x - 28 & -28 + 2x \\ \hline 3x &= -24 \\ \frac{3x}{3} &= \frac{-24}{3} \\ x &= -8 \end{aligned}$$

---

21. Solve for x: $\frac{x+1}{5} - 1 = 3$

Solve It

22. Solve for x: $\frac{2x}{3} - \frac{3x}{4} = 1$

Solve It

Part III: Stirring Up Solutions

23. Solve for y: $\dfrac{2(y+3)}{5} - 1 = \dfrac{3(y-3)}{4}$

Solve It

24. Solve for x: $\dfrac{x}{2} + \dfrac{x}{3} + \dfrac{x}{6} = 6$

Solve It

25. Solve for y: $\dfrac{4}{y} - \dfrac{6}{y} = 1$

Solve It

26. Solve for z: $\dfrac{1}{3z} - \dfrac{1}{2z} = \dfrac{1}{6}$

Solve It

Solving Proportions

A *proportion* is actually an equation with two fractions set equal to one another. The proportion $\frac{a}{b} = \frac{c}{d}$ has the following properties:

- The cross products are equal, $ad = bc$.
- If the proportion is true, then the *flip* of the proportion is also true, $\frac{b}{a} = \frac{d}{c}$.

You can solve equations involving proportions by cross-multiplying, flipping, or both. The flipping part of solving proportions usually occurs when you have the variable in the denominator and can do a quick solution by first flipping the proportion and then multiplying by the number in the denominator under the variable.

Q. Solve for x: $\frac{2x}{16} = \frac{3x+5}{28}$

A. $x = 10$

1. Cross-multiply.

   $2x \times 28 = 16(3x + 5)$
   $56x = 48x + 80$

2. Subtract $48x$ from each side.
3. Divide each side by 8.

   $56x = 48x + 80$
   $\underline{-48x -48x}$
   $8x = 80$
   $\frac{8x}{8} = \frac{80}{8}$
   $x = 10$

Q. Solve for x: $\frac{6}{10} = \frac{12}{x}$

A. $x = 20$

1. Flip the proportion to get $\frac{10}{6} = \frac{x}{12}$.
2. Multiply each side by 12 to get that $x = 20$.

   $12\left(\frac{10}{6}\right) = \left(\frac{x}{12}\right)12$
   $20 = x$

---

27. Solve for x: $\frac{x}{8} = \frac{9}{12}$

Solve It

28. Solve for y: $\frac{20}{y} = \frac{30}{33}$

Solve It

29. Solve for z: $\dfrac{z+4}{32} = \dfrac{35}{56}$

Solve It

30. Solve for y: $\dfrac{6}{27} = \dfrac{8}{2y+6}$

Solve It

31. Solve for x: $\dfrac{x+15}{10} = \dfrac{3x}{18}$

Solve It

32. Solve for x: $\dfrac{x+10}{x-4} = \dfrac{3x+15}{3x-20}$

Solve It

Working with Formulas

A *formula* is an equation that expresses some known relationship between given quantities. You use formulas to determine how much a dollar is worth when you go to another country. You use a formula to figure out how much paint to buy when redecorating your home.

The formulas covered in this chapter are all the types that involve variables raised to the first degree. Solving for one of variables in a formula is often advantageous if you have to repeat the same computation over and over again. In each of the examples and practice exercises, I tell you which variable to solve for or isolate. You use the same rules for solving — you'll just be getting an answer in terms of one or more variables rather than a single number.

Q. The formula for changing from degrees Celsius to degrees Fahrenheit is

$F = \frac{9}{5}C + 32$ where F is in degrees

Fahrenheit and C is in degrees Celsius. Solve for C.

A. $C = \frac{5}{9}(F - 32)$

Subtract 32 from each side, and then multiply each side by the reciprocal of $\frac{9}{5}$:

$$F = \frac{9}{5}C + 32$$
$$\underline{-32 \qquad -32}$$
$$F - 32 = \frac{9}{5}C$$
$$\frac{5}{9} \times (F - 32) = \frac{5}{9} \times \frac{9}{5}C$$
$$\frac{5}{9}(F - 32) = C$$

Don't bother distributing the $\frac{5}{9}$ over the terms in the parenthesis. The formula works just fine this way — it cuts down on computing with fractions.

Q. The formula for finding the area of a triangle is $A = \frac{1}{2}bh$. Solve for h.

A. $h = \frac{2A}{b}$

Divide each side by $\frac{1}{2}b$ or, to avoid a complex fraction, multiply each side by 2 and then divide each side by b:

$$A = \frac{1}{2}bh$$
$$2 \times A = 2 \times \frac{1}{2}bh$$
$$2A = bh$$
$$\frac{2A}{b} = \frac{bh}{b}$$
$$\frac{2A}{b} = h$$

33. The simple interest formula is $I = Prt$. Solve for t.

Solve It

34. The formula for the perimeter of an isosceles triangle is $P = 2s + b$. Solve for b.

Solve It

35. The formula for the perimeter of a rectangle is $P = 2(l + w)$. Solve for w.

Solve It

36. The formula for the area of a trapezoid is $A = \frac{1}{2}h(b_1 + b_2)$. Solve for b_2.

Solve It

Answers to Problems on Solving Linear Equations

This section provides the answers (in bold) to the practice problems in this chapter.

1 Solve for x: $x + 4 = 15$. The answer is $x = 11$.

$$\begin{array}{r} x + 4 = 15 \\ -4 -4 \\ \hline x = 11 \end{array}$$

2 Solve for y: $y - 2 = 11$. The answer is $y = 13$.

$$\begin{array}{r} y - 2 = 11 \\ +2 +2 \\ \hline y = 13 \end{array}$$

3 Solve for x: $5x + 3 = 4x - 1$. The answer is $x = -4$.

$$\begin{array}{r} 5x + 3 = 4x - 1 \\ -4x -4x \\ \hline x + 3 = -1 \\ -3 -3 \\ \hline x = -4 \end{array}$$

4 Solve for y: $2y + 9 + 6y - 8 = 4y + 5 + 3y - 11$. The answer is $y = -7$. By collecting like terms, you get

$$\begin{array}{r} 8y + 1 = 7y - 6 \\ -7y = -7y \\ \hline y + 1 = -6 \\ -1 -1 \\ \hline y = -7 \end{array}$$

5 Solve for x: $6x = 24$. The answer is $x = 4$.

$$6x = 24$$
$$\frac{6x}{6} = \frac{24}{6}$$
$$x = 4$$

6 Solve for y: $-4y = 20$. The answer is $y = -5$.

$$-4y = 20$$
$$\frac{-4y}{-4} = \frac{20}{-4}$$
$$y = -5$$

7 Solve for z: $\frac{z}{3} = 11$. The answer is $z = 33$.

$$\frac{z}{3} = 11$$
$$3\left(\frac{z}{3}\right) = (11)(3)$$
$$z = 33$$

8 Solve for w: $\frac{w}{-4} = -2$. The answer is $w = 8$.

$$\frac{w}{-4} = -2$$
$$(-4)\left(\frac{w}{-4}\right) = (-2)(-4)$$
$$w = 8$$

9 Solve for x: $3x - 4 = 5$. The answer is $x = 3$.

$$3x - 4 = 5$$
$$ + 4 +4$$
$$\overline{3x = 9}$$
$$\frac{3x}{3} = \frac{9}{3}$$
$$x = 3$$

10 Solve for y: $8 - \frac{y}{2} = 7$. The answer is $y = 2$.

$$8 - \frac{y}{2} = 7$$
$$-8 \phantom{- \frac{y}{2} =} -8$$
$$\overline{-\frac{y}{2} = -1}$$
$$(-2)\left(-\frac{y}{2}\right) = (-1)(-2)$$
$$y = 2$$

11 Solve for x: $5x - 3 = 8x + 9$. The answer is $-4 = x$.

$$5x - 3 = 8x + 9$$
$$-5x - 9 = -5x - 9$$
$$\overline{-12 = 3x}$$
$$\frac{-12}{3} = x$$
$$-4 = x$$

12 Solve for z: $\frac{z}{6} - 3 = z + 7$. The answer is $z = -12$.

$$\frac{z}{6} - 3 = z + 7$$
$$-\frac{z}{6} - 7 -\frac{z}{6} - 7$$
$$\overline{-10 = \frac{5}{6}z}$$

because $z - \frac{z}{6} = \frac{6z}{6} - \frac{z}{6} = \frac{5z}{6}$

$$(6)(-10) = \left(\frac{5}{6}z\right)(6)$$
$$-60 = 5z$$
$$\frac{5z}{5} = \frac{-60}{5}$$
$$z = -12$$

13 Solve for y: $4y + 16 - 3y = 7 + 3y$. The answer is $y = 9/2$. By collecting like terms,

$$y + 16 = 7 + 3y$$
$$\underline{-3y - 16 \quad -16 - 3y}$$
$$-2y \quad\quad = -9$$
$$\frac{-2y}{-2} = \frac{-9}{-2}$$
$$y = \frac{9}{2}$$

14 Solve for x: $\frac{3x}{4} - 2 = \frac{9x}{4} + 13$. The answer is $x = -10$.

$$\frac{3x}{4} - 2 = \frac{9x}{4} + 13$$
$$\underline{-\frac{9x}{4} + 2 \quad -\frac{9x}{4} + 2}$$
$$-\frac{6x}{4} \quad\quad = 15$$
$$-\frac{3x}{2} = 15$$
$$(2)\left(-\frac{3x}{2}\right) = (15)(2)$$
$$-3x = 30$$
$$\frac{-3x}{-3} = \frac{30}{-3}$$
$$x = -10$$

15 Solve for x: $3(x - 5) = 12$. The answer is $x = 9$.

$$3(x - 5) = 12$$
$$3x - 15 = 12$$
$$\underline{+15 \quad +15}$$
$$3x = 27$$
$$\frac{3x}{3} = \frac{27}{3}$$
$$x = 9$$

16 Solve for y: $4(y + 3) + 7 = 3$. The answer is $y = -4$.

$$4(y + 3) + 7 = 3$$
$$4y + 12 + 7 = 3$$
$$4y + 19 = 3$$
$$\underline{-19 \quad -19}$$
$$4y = -16$$
$$\frac{4y}{4} = \frac{-16}{4}$$
$$y = -4$$

17 Solve for x: $\frac{4x+1}{3} = x + 2$. The answer is $x = 5$.

$$\frac{4x+1}{3} = x + 2$$
$$(3)\left(\frac{4x+1}{3}\right) = (x+2)(3)$$
$$4x + 1 = 3x + 6$$
$$\underline{-3x - 1 \quad -3x - 1}$$
$$x = 5$$

18 Solve for y: $5(y-3) - 3(y+4) = 1 - 6(y-4)$. The answer is $y = 13/2$.

$$5(y-3) - 3(y+4) = 1 - 6(y-4)$$
$$5y - 15 - 3y - 12 = 1 - 6y + 24$$
$$2y - 27 = -6y + 25$$
$$\underline{+6y + 27 \quad +6y + 27}$$
$$8y = 52$$
$$\frac{8y}{8} = \frac{52}{8}$$
$$y = \frac{13}{2}$$

19 Solve for x: $x(3x+1) - 2 = 3x^2 - 5$. The answer is $x = -3$.

$$x(3x+1) - 2 = 3x^2 - 5$$
$$3x^2 + x - 2 = 3x^2 - 5$$
$$\underline{-3x^2 \quad\quad +2 \; -3x^2 + 2}$$
$$x = -3$$

20 Solve for x: $(x-3)(x+4) = (x+1)(x-2)$. The answer is $x = 5$.

$$(x-3)(x+4) = (x+1)(x-2)$$
$$x^2 + x - 12 = x^2 - x - 2$$
$$\underline{-x^2 + x + 12 \; -x^2 + x + 12}$$
$$2x = 10$$
$$\frac{2x}{2} = \frac{10}{2}$$
$$x = 5$$

21 Solve for x: $\frac{x+1}{5} - 1 = 3$. The answer is $x = 19$.

$$\frac{x+1}{5} - 1 = 3$$
$$\underline{\quad\quad +1 \;\; +1}$$
$$\frac{x+1}{5} = 4$$
$$(5)\left(\frac{x+1}{5}\right) = (4)(5)$$
$$x + 1 = 20$$
$$\underline{\;\; -1 \quad -1}$$
$$x = 19$$

22 Solve for x: $\frac{2x}{3} - \frac{3x}{4} = 1$. The answer is $x = -12$. Twelve is a common denominator, so

$$(12)\left(\frac{2x}{3}\right) - 12\left(\frac{3x}{4}\right) = 12(1)$$
$$8x - 9x = 12$$
$$-x = 12$$
$$\frac{-x}{-1} = \frac{12}{-1}$$
$$x = -12$$

Chapter 12: Putting It on the Line: Solving Linear Equations

23 Solve for y: $\dfrac{2(y+3)}{5} - 1 = \dfrac{3(y-3)}{4}$. The answer is **y = 7**. Twenty is a common denominator, so

$$20\left[\dfrac{2(y+3)}{5}\right] - 20(1) = 20\left[\dfrac{3(y-3)}{4}\right]$$
$$8(y+3) - 20 = 15(y-3)$$
$$8y + 24 - 20 = 15y - 45$$
$$8y + 4 = 15y - 45$$
$$\underline{-8y + 45 \quad -8y + 45}$$
$$49 = 7y$$
$$\dfrac{7y}{7} = \dfrac{49}{7}$$
$$y = 7$$

24 Solve for x: $\dfrac{x}{2} + \dfrac{x}{3} + \dfrac{x}{6} = 6$. The answer is **x = 6**. Six is a common denominator, so

$$6\left(\dfrac{x}{2}\right) + 6\left(\dfrac{x}{3}\right) + 6\left(\dfrac{x}{6}\right) = 6(6)$$
$$3x + 2x + x = 36$$
$$6x = 36$$
$$\dfrac{6x}{6} = \dfrac{36}{6}$$
$$x = 6$$

25 Solve for y: $\dfrac{4}{y} - \dfrac{6}{y} = 1$. The answer is **−2 = y**. Y is a common denominator, so

$$y\left(\dfrac{4}{y}\right) - y\left(\dfrac{6}{y}\right) = y(1)$$
$$4 - 6 = y$$
$$-2 = y$$

26 Solve for z: $\dfrac{1}{3z} - \dfrac{1}{2z} = \dfrac{1}{6}$. The answer is **−1 = z**. 6z is a common denominator, so

$$6z\left(\dfrac{1}{3z}\right) - 6z\left(\dfrac{1}{2z}\right) = 6z\left(\dfrac{1}{6}\right)$$
$$2 - 3 = z$$
$$-1 = z$$

27 Solve for x: $\dfrac{x}{8} = \dfrac{9}{12}$. The answer is **x = 6**. Multiply each side by 8:

$$\dfrac{x}{8} = \dfrac{9}{12}$$
$$8\left(\dfrac{x}{8}\right) = 8\left(\dfrac{9}{12}\right) = \dfrac{72}{12}$$
$$x = 6$$

28 Solve for y: $\dfrac{20}{y} = \dfrac{30}{33}$. The answer is **y = 22**. Flip to get $\dfrac{y}{20} = \dfrac{33}{30}$ and solve.

$$20\left(\dfrac{y}{20}\right) = 20\left(\dfrac{33}{30}\right)$$
$$y = 22$$

Part III: Stirring Up Solutions

29. Solve for z: $\frac{z+4}{32} = \frac{35}{56}$. The answer is $z = 16$. Reduce the fraction on the right by dividing the numerator and denominator by 7. Then multiply each side by 32:

$$\frac{z+4}{32} = \frac{35}{56} = \frac{\cancel{35}^5}{\cancel{56}_8} = \frac{5}{8}$$

$$32\left(\frac{z+4}{32}\right) = 32\left(\frac{5}{8}\right)$$

$$z + 4 = 20$$
$$-4 \quad -4$$
$$\overline{ z = 16}$$

30. Solve for y: $\frac{6}{27} = \frac{8}{2y+6}$. The answer is $y = 15$. Flip to get $\frac{27}{6} = \frac{2y+6}{8}$ and solve by reducing the fraction on the left and multiplying each side by 8:

$$8\left(\frac{\cancel{27}^9}{\cancel{6}_2}\right) = 8\left(\frac{2y+6}{8}\right)$$

$$36 = 2y + 6$$
$$-6 \quad\quad -6$$
$$\overline{30 = 2y}$$

$$\frac{2y}{2} = \frac{30}{2}$$

$$y = 15$$

31. Solve for x: $\frac{x+15}{10} = \frac{3x}{18}$. The answer is $x = \frac{45}{2}$.

$$\frac{x+15}{10} = \frac{3x}{18}$$

$$\frac{x+15}{10} = \frac{x}{6}$$

because $\frac{3}{18} = \frac{1}{6}$. Then cross-multiply and distribute the 6 and 10:

$$6(x+15) = 10(x)$$
$$6x + 90 = 10x$$
$$-6x \quad\quad -6x$$
$$\overline{90 = 4x}$$

$$\frac{4x}{4} = \frac{90}{4}$$

$$x = \frac{45}{2}$$

32. Solve for x: $\frac{x+10}{x-4} = \frac{3x+15}{3x-20}$. The answer is $x = 20$. First cross-multiply. The two binomials have to be multiplied on each side — use FOIL. Then simplify and solve for x:

$$(x+10)(3x-20) = (3x+15)(x-4)$$
$$3x^2 + 10x - 200 = 3x^2 + 3x - 60$$
$$-3x^2 \quad -3x + 200 \quad -3x^2 - 3x + 200$$
$$\overline{7x = 140}$$

$$\frac{7x}{7} = \frac{140}{7}$$

$$x = 20$$

Chapter 12: Putting It on the Line: Solving Linear Equations

33 The simple interest formula is $I = Prt$. Solve for t. The answer is $\dfrac{I}{Pr} = t$.

$$I = Prt$$
$$\dfrac{I}{Pr} = \dfrac{Prt}{Pr}$$
$$\dfrac{I}{Pr} = t$$

34 The formula for the perimeter of an isosceles triangle is $P = 2s + b$. Solve for b. The answer is **$P - 2s = b$**.

$$P = 2s + b$$
$$\underline{-2s \quad -2s}$$
$$P - 2s = b$$

35 The formula for the perimeter of a rectangle is $P = 2(l + w)$. Solve for w. The answer is $\dfrac{P}{2} - l = w$.

$$P = 2(l + w)$$
$$\dfrac{P}{2} = \dfrac{2(l+w)}{2}$$
$$\dfrac{P}{2} = l + w$$
$$\underline{\phantom{\dfrac{P}{2} =} -l \quad -l}$$
$$\dfrac{P}{2} - l = w$$

36 The formula for the area of a trapezoid is $A = \dfrac{1}{2}h(b_1 + b_2)$. Solve for b_2. The answer is $\dfrac{2A}{h} - b_1 = b_2$. First write $\dfrac{1}{2}h$ as $\dfrac{h}{2}$. Then multiply by the reciprocal.

$$A = \dfrac{1}{2}h(b_1 + b_2)$$
$$\dfrac{2}{h}(A) = \dfrac{2}{h}\left(\dfrac{h}{2}\right)(b_1 + b_2)$$
$$\dfrac{2A}{h} = b_1 + b_2$$
$$\underline{\phantom{\dfrac{2A}{h} =} -b_1 \quad -b_1}$$
$$\dfrac{2A}{h} - b_1 = b_2$$

Chapter 13

Solving Quadratic Equations

In This Chapter
- Understanding the square root rule
- Solving quadratic equations by factoring
- Using the quadratic formula
- Approximating answers
- Coping with the impossible

A *quadratic equation* is an equation that can be written in the form $ax^2 + bx + c = 0$ where b, c, or both b and c can be equal to 0, but a can't be equal to 0. The solutions of quadratic equations can be two numbers, one number, or no real number at all. (*Real numbers* are all the whole numbers, fractions, negatives and positives, radicals, and irrational decimals.)

With quadratic equations, you can have a *maximum* of two solutions, but you can have fewer solutions as well. For most solutions, you set the equation equal to 0, writing the terms in decreasing powers of the variable. *Note:* One exception is when you have just a squared term and a number, and you want to use the *square root rule*. I go into further detail about this exception in this chapter.

For now, strap on your boots and get ready to answer quadratic equations. This chapter offers you plenty of chances to get your feet wet.

Using the Square Root Rule

You can use the *square root rule* when a quadratic equation has just the squared term and a number — no term with the variable to the first degree. This rule says that if $x^2 = k$, then $x = \pm\sqrt{k}$, as long as k isn't a negative number. The tricky part here — what most people trip on — is in remembering to include *both* solutions by including the $\pm$ (or plus or minus, positive or negative, a space-saver) symbol.

Q. Use the square root rule to solve: $7x^2 = 28$

A. $x = \pm 2$ **(which is the same as $x = 2$ or $x = -2$).** First divide each side by 7 to get $x^2 = 4$. Then find the square root of 4 to get the answer.

Q. Use the square root rule to solve: $3y^2 - 75 = 0$

A. $y = \pm 5$ **(which is the same as $y = 5$ or $y = -5$).** Before using the square root rule, first add 75 to each side, and then divide each side by 3 to get $y^2 = 25$. Then find the square root of 25 to get the answer.

1. Use the square root rule to solve: $x^2 = 9$

Solve It

2. Use the square root rule to solve: $5y^2 = 80$

Solve It

3. Use the square root rule to solve: $z^2 - 100 = 0$

Solve It

4. Use the square root rule to solve: $20w^2 - 125 = 0$

Solve It

Solving by Factoring

The quickest and most efficient way of solving a quadratic equation is to factor it (if it can be factored) and then use the *multiplication property of zero* (MPZ) to solve for the solutions. The MPZ says that, if you multiply two or more factors together and they have a product of 0, then one or more of the factors must be equal to 0. You can use this property on the factored form of a quadratic equation to set the two linear factors equal to 0 and solve those simple equations for the value of the variable.

Sometimes you factor out the quadratic and get two identical factors. That's because the quadratic was a perfect square. You still get two answers, but they're the same number, which is called a *double root*. You can ignore any number that's a greatest common factor (GCF). A number set equal to zero won't give you an answer; you still just use the two binomial factors to solve for the solutions.

Q. Solve for x by factoring: $18x^2 + 21x - 60 = 0$

A. $x = \frac{4}{3}$ or $x = -\frac{5}{2}$. The factored form of the original equation is $3(6x^2 + 7x - 20) = 3(3x - 4)(2x + 5) = 0$. Setting the 3 equal to 0 is strictly correct by the multiplication property of zero, but it doesn't make any sense because the number 3 is never equal to 0. Only the variables can take on a value to make the factor equal to 0. Therefore you need to set the other two factors equal to 0 and solve those equations to give you the answers.

Q. Solve for x by factoring: $4x^2 - 48x = 0$

A. $x = 0$ or $x = 12$. You don't have to have the constant term (it's technically 0 in this case) in order for this process to work. Just factor out $4x$ from each term in the equation. Setting each factor equal to 0, you get

$$4x(x - 12) = 0$$
$$4x = 0, \; x = 0$$
$$x - 12 = 0, \; x = 12$$

Don't forget to set that first factor equal to 0, $x = 0$.

5. Solve for x by factoring: $x^2 - 2x - 15 = 0$

Solve It

6. Solve for x by factoring: $3x^2 - 25x + 28 = 0$

Solve It

7. Solve for y by factoring: $4y^2 - 9 = 0$

Solve It

8. Solve for z by factoring: $z^2 + 64 = 16z$

Solve It

9. Solve for y by factoring: $y^2 + 21y = 0$

Solve It

10. Solve for x by factoring: $12x^2 = 24x$

Solve It

11. Solve for z by factoring: $15z^2 + 14z = 0$

Solve It

12. Solve for y by factoring: $\frac{1}{4}y^2 = \frac{2}{3}y$

Solve It

Using the Quadratic Formula

You can use the quadratic formula to solve a quadratic equation whether the terms can be factored or not. Factoring and using the MPZ is always easier, but you'll encounter some equations that can't be factored and some equations that have numbers too large to factor in your head.

If a quadratic equation appears in its standard form, $ax^2 + bx + c = 0$, then the quadratic formula gives you the solutions to that equation. You can find the solutions using the following formula: $x = \dfrac{-b \pm \sqrt{b^2 - 4ac}}{2a}$

You can use the quadratic formula whether the equation can be factored or not. If the equation can be factored, then the number under the *radical* (the square root symbol) is a perfect square, and there's no radical in the solution. If the equation isn't factorable, then your answer — a perfectly good one, mind you — has radicals in it. If the number under the radical turns out to be negative, then that equation doesn't have a solution. If you want to know more about this situation, then look at the "Dealing with Impossible Answers" section later in this chapter.

Part III: Stirring Up Solutions

Q. Use the quadratic formula to solve: $2x^2 + 11x - 21 = 0$ with $a = 2$, $b = 11$, and $c = -21$

A. $x = \frac{3}{2}$ **or** -7. Fill in the formula and simplify to get

$$x = \frac{-11 \pm \sqrt{11^2 - 4(2)(-21)}}{2(2)}$$

$$= \frac{-11 \pm \sqrt{121 + 168}}{4}$$

$$= \frac{-11 \pm \sqrt{289}}{4} = \frac{-11 \pm 17}{4}$$

$$x = \frac{-11 + 17}{4} = \frac{6}{4} = \frac{3}{2} \text{ or}$$

$$x = \frac{-11 - 17}{4} = \frac{-28}{4} = -7$$

Because the 289 under the radical is a perfect square, you could have solved it by factoring.

Q. Use the quadratic formula to solve: $x^2 - 2x - 10 = 0$ with $a = 1$, $b = -2$, and $c = -10$

A. $x = 1 + \sqrt{11}$ **or** $x = 1 - \sqrt{11}$. Fill in the formula and simplify to get

$$x = \frac{2 \pm \sqrt{(-2)^2 - 4(1)(-10)}}{2(1)} = \frac{2 \pm \sqrt{4 + 40}}{2}$$

$$= \frac{2 \pm \sqrt{44}}{2} = \frac{2 \pm \sqrt{4(11)}}{2} = \frac{2 \pm 2\sqrt{11}}{2}$$

$$= 1 \pm \sqrt{11}$$

$$x = 1 + \sqrt{11} \text{ or } x = 1 - \sqrt{11}$$

You have to use the quadratic formula to solve this one.

---

13. Use the quadratic formula to solve: $x^2 - 5x - 6 = 0$

Solve It

14. Use the quadratic formula to solve: $6x^2 + x = 12$

Solve It

15. Use the quadratic formula to solve:
$x^2 - 4x - 6 = 0$

Solve It

16. Use the quadratic formula to solve:
$2x^2 + 9x = 2$

Solve It

17. Use the quadratic formula to solve:
$3x^2 - 5x = 0$. ***Hint:*** When there's no constant term, let $c = 0$.

Solve It

18. Use the quadratic formula to solve:
$4x^2 - 25 = 0$. ***Hint:*** Let $b = 0$.

Solve It

Part III: Stirring Up Solutions

Estimating Answers

When solutions for quadratic equations contain square roots of numbers that aren't perfect squares, you have two options.

- You can write the answers with the radicals (the *exact* answers).
- You can find a decimal estimate of the answer. The decimal values for these radical numbers continue on forever without repeating.

You just have to choose what accuracy you want the answer to have.

For instance, $\sqrt{2}$ has a value between 1 and 2. Rounded to one decimal place, it's equal to 1.4, and rounded to two decimal places it's equal to 1.41. If you're drawing a graph and have $\sqrt{2}$ as a coordinate of a point, then one decimal place is enough for most graphs. If you're working on something that needs more precision, then carry out the decimal values farther.

You can find the decimal approximations for the square roots of numbers by using a hand-held calculator. Most scientific calculators give you at least six decimal places.

Check out the values of two square roots rounded to six decimal places: $\sqrt{2} \approx 1.414214$ and $\sqrt{3} \approx 1.732051$. Use these two values in the following examples and practice problems:

Q. Estimate $\sqrt{48}$ to three decimal places.

A. **6.928.** Simplify the radical, and then substitute the value of the radical to three decimal places:

$$\sqrt{48} = \sqrt{16(3)} = 4\sqrt{3} \approx 4(1.732) = 6.928$$

Q. Simplify $\dfrac{1 \pm \sqrt{27}}{2}$.

A. **3.098075 and −2.098075.** Simplifying the results of the quadratic formula can involve this same type of estimating. First simplify the radical, and then substitute in the decimal value. I use five decimal places this time. You compute the two different answers with that decimal value.

$$\dfrac{1 \pm \sqrt{27}}{2} = \dfrac{1 \pm \sqrt{9(3)}}{2}$$
$$= \dfrac{1 \pm 3\sqrt{3}}{2}$$
$$\approx \dfrac{1 \pm 3(1.73205)}{2}$$
$$= \dfrac{1 \pm 5.19615}{2}$$

$\dfrac{1 + 5.19615}{2} = 3.098075$ and
$\dfrac{1 - 5.19615}{2} = -2.098075$

If you round back to five decimal points, you round it up to the even digit (in this case 8).

Chapter 13: Solving Quadratic Equations

19. Estimate $\sqrt{12}$ to three decimal places.

Solve It

20. Estimate $\sqrt{50}$ to four decimal places.

Solve It

21. Find both values to three decimal places:
$$\frac{5 \pm \sqrt{18}}{2}$$

Solve It

22. Find both values to four decimal places:
$$\frac{-3 \pm \sqrt{75}}{3}$$

Solve It

Dealing with Impossible Answers

The square root of a negative number doesn't exist — as far as real numbers are concerned. When you encounter the square root of a negative number when solving quadratic equations, it means that the equation doesn't have a solution.

If a negative under the radical shows up when you're solving a quadratic equation, that tells you there are no *x*-intercepts for the graph of the parabola. Refer to Chapter 21 for more on parabolas.

You can report your answers when you end up with negatives under the radical by using *imaginary numbers* (or numbers that are indicated with an *i* to show that they aren't real). With imaginary numbers, you let $\sqrt{-1} = i$. Then you can simplify the radical using the imaginary number as a factor.

Don't worry if you're slightly confused about imaginary numbers. This concept is more advanced than most of the algebra problems you'll encounter in this book and during Algebra I. For now, just remember that $\sqrt{-1} = i$ allows you to make answers more complete — imaginary numbers can allow you to finish the factoring.

Q. Rewrite $\sqrt{-9}$ using imaginary numbers.

A. **3*i*.** Split up the value under the radical into two factors:

$\sqrt{-9} = \sqrt{(-1)9} = \sqrt{-1}\sqrt{9} = i3$ or $3i$

Q. Rewrite $\sqrt{-48}$ using imaginary numbers.

A. **$4i\sqrt{3}$.** You can write the value under the radical as the product of three factors:

$\sqrt{-48} = \sqrt{(-1)(16)3}$
$= \sqrt{-1}\sqrt{16}\sqrt{3}$
$= i4\sqrt{3}$ or $4i\sqrt{3}$

23. Rewrite $\sqrt{-4}$ as a product with a factor of *i*.

Solve It

24. Rewrite $\sqrt{-96}$ as a product with a factor of *i*.

Solve It

Answers to Problems on Solving Quadratic Equations

This section provides the answers (in bold) to the practice problems in this chapter.

1 Use the square root rule to solve: $x^2 = 9$. The answer is $x = \pm 3$.

2 Use the square root rule to solve: $5y^2 = 80$. The answer is $y = \pm 4$.

$5y^2 = 80$

$\dfrac{5y^2}{5} = \dfrac{80}{5}$

$y^2 = 16$

$y = \pm 4$

3 Use the square root rule to solve: $z^2 - 100 = 0$. The answer is $z = \pm 10$.

$z^2 - 100 = 0$

$+100 +100$

$z^2 = 100$

$z = \pm 10$

4 Use the square root rule to solve: $20w^2 - 125 = 0$. The answer is $w = \pm\dfrac{5}{2}$.

$20w^2 - 125 = 0$

$+125 +125$

$20w^2 = 125$

$\dfrac{20w^2}{20} = \dfrac{125}{20}$

$w^2 = \dfrac{25}{4}$

$w = \pm\dfrac{5}{2}$

5 Solve for x by factoring: $x^2 - 2x - 15 = 0$. The answer is $x = 5$ or $x = -3$.

$x^2 - 2x - 15 = 0$

$(x - 5)(x + 3) = 0$

$x - 5 = 0, \ x = 5$

or $x + 3 = 0, \ x = -3$

$x = 5, -3$

6 Solve for x by factoring: $3x^2 - 25x + 28 = 0$. The answer is $x = \frac{4}{3}$ or $x = 7$.

$3x^2 - 25x + 28 = 0$

$(3x - 4)(x - 7) = 0$

$3x - 4 = 0, \ 3x = 4, \ x = \dfrac{4}{3}$

or $x - 7 = 0, \ x = 7$

$x = \dfrac{4}{3}, 7$

7 Solve for y by factoring: $4y^2 - 9 = 0$. The answer is $y = \pm\frac{3}{2}$.

$$4y^2 - 9 = 0$$
$$(2y+3)(2y-3) = 0$$
$$2y+3 = 0,\ 2y = -3,\ y = -\frac{3}{2}$$
$$\text{or}\ \ 2y-3 = 0,\ 2y = 3,\ y = \frac{3}{2}$$
$$y = \pm\frac{3}{2}$$

8 Solve for z by factoring: $z^2 + 64 = 16z$. The answer is $z = 8$ or $z = 8$, which is a double root.

$$z^2 + 64 = 16z$$
$$z^2 - 16z + 64 = 0$$
$$(z-8)(z-8) = 0$$
$$z - 8 = 0,\ z = 8$$
$$\text{or}\ \ z - 8 = 0,\ z = 8$$

9 Solve for y by factoring: $y^2 + 21y = 0$. The answer is $y = 0$ or $y = -21$.

$$y^2 + 21y = 0$$
$$y(y+21) = 0$$
$$y = 0$$
$$\text{or}\ \ y + 21 = 0,\ y = -21$$
$$y = 0, -21$$

10 Solve for x by factoring: $12x^2 = 24x$. The answer is $x = 0$ or $x = 2$.

$$12x^2 = 24x$$
$$12x^2 - 24x = 0$$
$$12x(x-2) = 0$$
$$12x = 0,\ x = 0$$
$$\text{or}\ \ x - 2 = 0,\ x = 2$$
$$x = 0, 2$$

11 Solve for z by factoring: $15z^2 + 14z = 0$. The answer is $z = 0$ or $z = -\frac{14}{15}$.

$$15z^2 + 14z = 0$$
$$z(15z + 14) = 0$$
$$z = 0$$
$$\text{or}\ \ 15z + 14 = 0,\ 15z = -14,\ z = -\frac{14}{15}$$
$$z = 0, -\frac{14}{15}$$

12 Solve for y by factoring: $\frac{1}{4}y^2 = \frac{2}{3}y$. The answer is $y = 0$ or $y = \frac{8}{3}$.

$\frac{1}{4}y^2 = \frac{2}{3}y$ has 12 as a common denominator, so

$$12\left(\frac{1}{4}y^2\right) = 12\left(\frac{2}{3}y\right)$$
$$3y^2 = 8y$$
$$3y^2 - 8y = 0$$

Chapter 13: Solving Quadratic Equations **165**

If you divide each side by y, you lose a solution:

$y(3y - 8) = 0$

$y = 0$

or $3y - 8 = 0$, $3y = 8$, $y = \frac{8}{3}$

$y = 0, \frac{8}{3}$

13 Use the quadratic formula to solve: $x^2 - 5x - 6 = 0$. The answer is $x = 6$ or $x = -1$.

$x^2 - 5x - 6 = 0$

$x = \dfrac{-(-5) \pm \sqrt{(-5)^2 - 4(1)(-6)}}{2(1)} = \dfrac{5 \pm \sqrt{25 + 24}}{2}$

$= \dfrac{5 \pm \sqrt{49}}{2} = \dfrac{5 \pm 7}{2}$

$= \dfrac{5 + 7}{2} = \dfrac{12}{2} = 6$ or $x = \dfrac{5 - 7}{2} = \dfrac{-2}{2} = -1$

$x = 6, -1$

This answer isn't unexpected. When a quadratic can be factored, as this one can, $x^2 - 5x^2 - 6 = (x - 6)(x + 1)$, then the answers comes out as rational numbers.

14 Use the quadratic formula to solve: $6x^2 + x = 12$. The answer is $x = \frac{4}{3}$ or $x = -\frac{3}{2}$.

$6x^2 + x = 12$

$6x^2 + x - 12 = 0$

$x = \dfrac{-(1) \pm \sqrt{(1)^2 - 4(6)(-12)}}{2(6)} = \dfrac{-1 \pm \sqrt{1 + 288}}{12}$

$= \dfrac{-1 \pm \sqrt{289}}{12} = \dfrac{-1 \pm 17}{12}$

$x = \dfrac{-1 + 17}{12} = \dfrac{16}{12} = \dfrac{4}{3}$ or $x = \dfrac{-1 - 17}{12} = \dfrac{-18}{12} = -\dfrac{3}{2}$

$x = \dfrac{4}{3}, -\dfrac{3}{2}$

15 Use the quadratic formula to solve: $x^2 - 4x - 6 = 0$. The answer is $2 \pm \sqrt{10}$.

$x^2 - 4x - 6 = 0$

$x = \dfrac{-(-4) \pm \sqrt{(-4)^2 - 4(1)(-6)}}{2(1)} = \dfrac{4 \pm \sqrt{16 + 24}}{2}$

$= \dfrac{4 \pm \sqrt{40}}{2} = \dfrac{4 \pm 2\sqrt{10}}{2} = 2 \pm \sqrt{10}$

16 Use the quadratic formula to solve: $2x^2 + 9x = 2$. The answer is $\dfrac{-9 \pm \sqrt{97}}{4}$.

$2x^2 + 9x = 2$

$2x^2 + 9x - 2 = 0$

$x = \dfrac{-(9) \pm \sqrt{(9)^2 - 4(2)(-2)}}{2(2)} = \dfrac{-9 \pm \sqrt{81 + 16}}{4}$

$= \dfrac{-9 \pm \sqrt{97}}{4}$

17 Use the quadratic formula to solve: $3x^2 - 5x = 0$. The answer is $x = 5/3$ or $x = 0$.

$3x^2 - 5x = 0$

$$x = \frac{-(-5) \pm \sqrt{(-5)^2 - 4(3)(0)}}{2(3)} = \frac{5 \pm \sqrt{25}}{6} = \frac{5 \pm 5}{6}$$

$$x = \frac{5+5}{6} = \frac{10}{6} = \frac{5}{3} \text{ or } x = \frac{5-5}{6} = \frac{0}{6} = 0$$

$$x = \frac{5}{3}, 0$$

Note: You could have factored the quadratic: $x(3x - 5) = 3x^2 - 5x = 0$

18 Use the quadratic formula to solve: $4x^2 - 25 = 0$. The answer is $x = \pm\frac{5}{2}$.

$4x^2 - 25 = 0$

$$x = \frac{-(0) \pm \sqrt{(0)^2 - 4(4)(-25)}}{2(4)} = \frac{\pm\sqrt{16 \times 25}}{8}$$

$$= \pm\frac{4 \times 5}{8} = \pm\frac{5}{2}$$

19 Estimate $\sqrt{12}$ to three decimal places. The answer is **3.464**.

$\sqrt{12} = \sqrt{4 \times 3} = 2\sqrt{3} \approx 2(1.732051) \approx 3.464$

20 Estimate $\sqrt{50}$ to four decimal places. The answer is **7.0711**.

$\sqrt{50} = \sqrt{25 \times 2} = 5\sqrt{2} \approx 5(1.414214) \approx 7.0711$

21 Find both values to three decimal places: $\frac{5 \pm \sqrt{18}}{2}$. The answer is **4.621 or .379**.

$$\frac{5 \pm \sqrt{18}}{2} = \frac{5 \pm \sqrt{9 \times 2}}{2} = \frac{5 \pm 3\sqrt{2}}{2} \approx \frac{5 \pm 3(1.414214)}{2} = \frac{5 \pm 4.242642}{2}$$

$$\frac{5 + 4.242642}{2} = \frac{9.242642}{2} \approx 4.621 \text{ or } \frac{5 - 4.242642}{2} \approx .379$$

22 Find both values to four decimal places: $\frac{-3 \pm \sqrt{75}}{3}$. The answer is **1.8868 or –3.8868**.

$$\frac{-3 \pm \sqrt{75}}{3} = \frac{-3 \pm \sqrt{25 \times 3}}{3} = \frac{-3 \pm 5\sqrt{3}}{3}$$

$$\frac{-3 + 5\sqrt{3}}{3} \approx \frac{-3 + 5(1.732051)}{3} = \frac{5.660255}{3} \approx 1.8868$$

$$\frac{-3 - 5\sqrt{3}}{3} \approx \frac{-3 - 5(1.732051)}{3} = \frac{-11.660255}{3} \approx -3.8868$$

23 Rewrite $\sqrt{-4}$ as a product with a factor of i. The answer is **$2i$**.

$\sqrt{-4} = \sqrt{(-1)(4)} = \sqrt{-1}\sqrt{4} = i(2) = 2i$

24 Rewrite $\sqrt{-96}$ as a product with a factor of i. The answer is **$4i\sqrt{6}$**.

$\sqrt{-96} = \sqrt{(-1)(16)(6)} = \sqrt{-1}(4)(\sqrt{6}) = i(4\sqrt{6}) = 4i\sqrt{6}$

Chapter 14
Yielding to Higher Powers

In This Chapter

▶ Counting up the possible number of roots
▶ Making educated guesses about solutions
▶ Using factoring to solve equations
▶ Recognizing quadratic-like equations

*P*olynomial equations have terms separated by addition and subtraction and have exponents that are always whole numbers. These equations are set equal to zero and then solved for the values that make them true statements. You can use the solutions to polynomial equations to solve problems in calculus, algebra, and other mathematical areas. You graph the solutions by showing where the curve intersects with the x-axis — either crossing it or just touching it at that point. Rather than just taking some wild guesses as to what the solutions might be, you can utilize some of the available techniques that help you make more reasonable guesses as to what the solutions are.

This chapter provides several examples of these techniques and gives you ample opportunities to try them out.

Determining the Number of Possible Roots

A rule known as *Descartes' rule of sign* allows you to estimate the number of real roots that a polynomial equation might have. The *real roots* are the real numbers that work in the equation. They may be the answer to a story problem or the place where the graph cuts through the x-axis. This rule doesn't tell you for sure how many roots — it just tells you the maximum number that there could be (if it's less than the maximum number of roots, then it's less than that by two or four or six, and so on). First, write the polynomial in decreasing powers of the variable. To count the possible number of *negative roots,* replace all the x's with negative x's. Simplify the terms, and count how many times the signs change.

Part III: Stirring Up Solutions

Q. How many possible real roots are there in $3x^5 + 5x^4 - x^3 + 2x^2 - x + 4 = 0$?

A. At most four positive and one negative. The sign changes from positive to negative to positive to negative to positive. That's four changes in sign, so you have a maximum of four positive real roots. If it doesn't have four, then it could have two. If it doesn't have two, then it has none. You step down by twos. Now count the number of possible negative real roots in that same polynomial by replacing all the x's with negative x's and counting the number of sign changes:

$$3(-x)^5 + 5(-x)^4 - (-x)^3 + 2(-x)^2 - (-x) + 4 = 0$$
$$= -3x^5 + 5x^4 + x^3 + x^2 + x + 4 = 0$$

This version only has one sign change — from negative to positive, which means that it has one negative real root. You can't go down by two from that, so it's the only choice.

Q. How many possible real roots are there in $6x^4 + 5x^3 + 3x^2 + 2x - 1 = 0$?

A. At most one positive and three or one negative. Count the number of sign changes in the original equation. It has only one sign change, so there's one positive real root. Change the function by replacing all the x's with negative x's and count the changes in sign:

$$6(-x)^4 + 5(-x)^3 + 3(-x)^2 + 2(-x) - 1 = 0$$
$$= 6x^4 - 5x^3 + 3x^2 - 2x - 1 = 0$$

It has three sign changes, which means that it has three or one negative real roots.

1. Count the number of possible positive and negative real roots in $x^5 - x^3 + 8x^2 - 8 = 0$.

Solve It

2. Count the number of possible positive and negative real roots in $8x^5 - 25x^4 - x^2 + 25 = 0$.

Solve It

Applying the Rational Root Theorem

In the previous section, you discover that *Descartes' rule of sign* counts the possible number of *real* roots. Now you need to understand Descartes' rule in order to figure out just what those roots are.

Real numbers can be either rational or irrational. (*Rational* numbers are numbers that have a fractional equivalent — they can be written as a fraction. *Irrational* numbers can't be written as a fraction; they have decimal values that never repeat and never end.) You can write a rational number as a fraction, and you end up writing fractions to find the rational roots.

The *Rational root theorem* says that if you have a polynomial equation written in the form $a_n x^n + a_{n-1} x^{n-1} + a_{n-2} x^{n-2} + \ldots + a_1 x^1 + a_0 = 0$, then you can make a list of all the possible rational roots by looking at the first term and the last term. Any rational roots must be able to be written as a fraction with a factor of the *constant* (the last term or a_0) in the numerator of the fraction and a factor of the lead coefficient (a_n) in the denominator.

For example, in the equation $4x^4 - 3x^3 + 5x^2 + 9x - 3 = 0$, the factors of the constant are +3, −3, +1, −1 and the factors of the lead coefficient are +4, −4, +2, −2, +1, −1. The following list includes all the ways that you can create a fraction with a factor of the constant in the numerator and a factor of the lead coefficient in the denominator:

$$\frac{+3}{+4}, \frac{+3}{-4}, \frac{+3}{+2}, \frac{+3}{-2}, \frac{+3}{+1}, \frac{+3}{-1}, \frac{-3}{+4}, \frac{-3}{-4}, \frac{-3}{+2}, \frac{-3}{-2}, \frac{-3}{+1}, \frac{-3}{-1},$$
$$\frac{+1}{+4}, \frac{+1}{-4}, \frac{+1}{+2}, \frac{+1}{-2}, \frac{+1}{+1}, \frac{+1}{-1}, \frac{-1}{+4}, \frac{-1}{-4}, \frac{-1}{+2}, \frac{-1}{-2}, \frac{-1}{+1}, \frac{-1}{-1}$$

You probably see several repeats when these fractions are simplified. For instance, $\frac{+3}{-4}$ and $\frac{-3}{+4}$ have the same value. A new, simplified, condensed list is $\pm\frac{3}{4}, \pm\frac{3}{2}, \pm 3, \pm\frac{1}{4}, \pm\frac{1}{2}, \pm 1$.

Although this new list has twelve candidates for solutions to the equation, it's relatively short when you're trying to run through all the possibilities. Many of the polynomials start out with a 1 as the coefficient of the first term, which is great news when you're writing your list, because that means the only rational numbers you're considering are whole numbers — the denominators are 1.

Q. Determine all the possible rational solutions of this equation:
$2x^6 - 4x^3 + 5x^2 + x - 30 = 0$

A. $\pm 30, \pm 15, \pm\frac{15}{2}, \pm 10, \pm 6, \pm 5, \pm\frac{5}{2}, \pm 3,$ $\pm\frac{3}{2}, \pm 2, \pm 1, \pm\frac{1}{2}$. The factors of the constant are $\pm 30, \pm 15, \pm 10, \pm 6, \pm 5, \pm 3, \pm 2, \pm 1$ and the factors of the lead coefficient are $\pm 2, \pm 1$. You create the list of all the numbers that could be considered for roots of the equation by dividing each of the factors of the constant by the factors of the lead coefficient.

Q. What's the capital of Australia?

A. **Canberra**. Did you guess Melbourne or even Sydney? Try again. You're thinking why I asked you. Well, if you ever decide to go Down Under, you may need to know where to look up the Australian prime minister to tell him "G'day mate!"

3. List all the possible rational roots of $2x^4 - 3x^3 - 54x + 81 = 0$.

Solve It

4. List all the possible rational roots of $8x^5 - 25x^3 - x^2 + 25 = 0$.

Solve It

Using the Factor/Root Theorem

Algebra has a theorem that says if the binomial $x - c$ is a factor of a polynomial, then c is a root or solution of the polynomial. You may say, "Okay, so what?" Well, this property means that you can use the very efficient method of synthetic division to solve for solutions to these polynomial equations.

Just remember that you can use synthetic division to try out all those rational numbers that are possibilities for roots of a polynomial. (See Chapter 8 for more on synthetic division.) If $x - c$ is a factor (and c is a root), then you don't have a remainder (the remainder is 0) when you perform synthetic division.

Q. Check to see if the number 2 is a root of the following polynomial:

$$x^6 - 6x^5 + 8x^4 + 2x^3 - x^2 - 7x + 2 = 0$$

A. **Yes, it's a root.** Use the 2 and the coefficients of the polynomial in a synthetic division problem:

$$\underline{2|}\ \ 1\ \ -6\ \ \ \ 8\ \ \ \ 2\ \ -1\ \ -7\ \ \ \ 2$$
$$\phantom{\underline{2|}\ \ 1}\ \ \ \ \ \ \ \ 2\ \ -8\ \ \ \ 0\ \ \ \ 4\ \ \ \ 6\ \ -2$$
$$\phantom{\underline{2|}\ \ }\ \ 1\ \ -4\ \ \ \ 0\ \ \ \ 2\ \ \ \ 3\ \ -1\ \ \ \ 0$$

The remainder is 0, so $x - 2$ is a factor and 2 is a root or solution. The quotient of this division is $x^5 - 4x^4 + 2x^2 + 3x - 1$, which you write using the coefficients along the bottom. Make sure you start with a variable that's one degree lower than the one that was divided into. This new polynomial ends in a -1 and has a lead coefficient of 1, so the only possible solutions are 1 or -1.

Q. Check to see if 1 or -1 is a solution of the new equation.

A. **Neither is a solution.** First, try 1:

$$\underline{1|}\ \ 1\ \ -4\ \ \ \ 0\ \ \ \ 2\ \ \ \ 3\ \ -1$$
$$\phantom{\underline{1|}\ \ 1}\ \ \ \ \ \ \ \ 1\ \ -3\ \ -3\ \ -1\ \ \ \ 2$$
$$\phantom{\underline{1|}\ \ }\ \ 1\ \ -3\ \ -3\ \ -1\ \ \ \ 2\ \ \ \ 1$$

That one didn't work; the remainder isn't 0. Now, try -1:

$$\underline{-1|}\ \ 1\ \ -4\ \ \ \ 0\ \ \ \ 2\ \ \ \ 3\ \ -1$$
$$\phantom{\underline{-1|}\ \ 1}\ \ \ \ \ \ -1\ \ \ \ 5\ \ -5\ \ \ \ 3\ \ -6$$
$$\phantom{\underline{-1|}\ \ }\ \ 1\ \ -5\ \ \ \ 5\ \ -3\ \ \ \ 6\ \ -7$$

It doesn't work, either. 2 is the only real solution of the original equation.

---

5. Check to see if -3 is a root of $x^4 - 10x^2 + 9 = 0$.

Solve It

6. Check to see if 8 is a root of $x^5 - x^3 + 8x^2 - 8 = 0$.

Solve It

7. Check to see if $\frac{3}{2}$ is a root of
$2x^4 - 3x^3 - 54x + 81 = 0$.

Solve It

8. Check to see if -1 is a root of
$8x^5 - 25x^3 - x^2 + 25 = 0$.

Solve It

Solving By Factoring

When determining the solutions for polynomials, many techniques are available to help you determine what those solutions are — if there are any. One method that is usually the quickest, though, is factoring and using the *multiplication property of zero (MPZ)*. (Check out Chapter 13 for more ways to use the MPZ.) Not all polynomials lend themselves to factoring, but, when they do, using this method is to your advantage.

Q. Find the real solutions of $x^4 - 81 = 0$.

A. $x = 3$ **or** $x = -3$. Do you recognize that the two numbers are both perfect squares? Factoring the binomial into the sum and difference of the roots, you get $(x^2 - 9)(x^2 + 9) = 0$. The first factor of this factored form is also the difference of perfect squares. Factoring again, you get $(x - 3)(x + 3)(x^2 + 9) = 0$. Now, to use the MPZ, set the first factor equal to 0 to get $x - 3 = 0, x = 3$. Set the second factor equal to 0, $x + 3 = 0, x = -3$. The last factor doesn't cooperate: $x^2 + 9 = 0, x^2 = -9$. A perfect square can't be negative, so this factor has no solution. If you go back to the original equation and use *Descartes' rule of sign*, (see "Determining The Number of Possible Roots" earlier in this chapter), you see that it has one real positive root and one real negative root, which just confirms that prediction.

Q. Find the real solutions of
$x^4 + 2x^3 - 125x - 250 = 0$.

A. $x = -2$ **or** $x = 5$. You can factor by grouping to get $x^3(x + 2) - 125(x + 2) = (x + 2)(x^3 - 125) = 0$. The second factor is the difference of perfect cubes, which factors $(x + 2)(x - 5)(x^2 + 5x + 25) = 0$. The trinomial factor in this factorization never factors any more, so only the first two factors yield solutions: $x + 2 = 0, x = -2$ and $x - 5 = 0, x = 5$

9. Solve by factoring: $x^4 - 16 = 0$

Solve It

10. Solve by factoring: $x^8 - x^2 = 0$

Solve It

11. Solve by factoring: $x^3 + 5x^2 - 16x - 80 = 0$

Solve It

12. Solve by factoring: $x^6 - 9x^4 - 16x^2 + 144 = 0$

Solve It

Solving Powers That Are Quadratic-Like

A special classification of equations, called *quadratic-like,* lends itself to solving by using the patterns found in quadratic equations that can be factored. These equations have three terms:

- A variable term with a particular power
- Another variable with a power half that of the first term
- A constant

A way to generalize these characteristics is with the equation $ax^{2n} + bx^n + c = 0$. When you see this pattern, you can try to factor the expression into the product of two binomials and then set the binomials equal to 0 to solve for an answer or answers.

Chapter 14: Yielding to Higher Powers

Q. Solve for x: $x^8 - 17x^4 + 16 = 0$

A. $x = 1$ or $x = -1$ or $x = 2$ or $x = -2$

1. Factor the expression into $(x^4 - 1)(x^4 - 16) = 0$.

2. Set the first factor equal to 0.

   You can factor $x^4 - 1 = 0$ again, as the difference of two squares: $(x^2 - 1)(x^2 + 1) = 0$. You can solve the first factor by using the square root rule, $x^2 = 1$, $x = \pm 1$. The second factor can't be factored and has no real solutions.

3. Set the second factor equal to 0 and factor.

   You get $x^4 - 16 = (x^2 - 4)(x^2 + 4) = 0$.

4. Take the first of those factors, set it equal to 0, and apply the square root rule.

   $x^2 - 4 = 0$, $x^2 = 4$, $x = \pm 2$

   Again, the other factor has no real solutions. The four solutions of the original equation are $x = -1, +1, -2$ and $+2$.

Q. Solve for y: $y^{2/3} + 5y^{1/3} + 6 = 0$

A. $y = -8$ or $y = -27$. This example involves fractional exponents. Notice that the power on the first variable is twice that of the second.

1. Factor the expression into $(y^{1/3} + 2)(y^{1/3} + 3) = 0$.

2. Take the first factor and set it equal to 0.

   $y^{1/3} + 2 = 0$

3. Add –2 to each side and cube each side of the equation.

   $y^{1/3} = -2$, $\left(y^{1/3}\right)^3 = (-2)^3$, $y = -8$

4. Follow the same steps with the other factor.

   $y^{1/3} + 3 = 0$, $y^{1/3} = -3$, $\left(y^{1/3}\right)^3 = (-3)^3$, $y = -27$

---

13. Solve the equation: $x^4 - 13x^2 + 36 = 0$

Solve It

14. Solve the equation: $x^{10} - 31x^5 - 32 = 0$

Solve It

---

15. Solve the equation: $y^{4/3} - 9y^{2/3} + 8 = 0$

Solve It

16. Solve the equation: $z^{-2} + z^{-1} - 12 = 0$

Solve It

Answers to Problems on Solving Higher Power Equations

This section provides the answers (in bold) to the practice problems in this chapter.

1 Count the number of possible positive and negative real roots in $x^5 - x^3 + 8x^2 - 8 = 0$. The answer: **Three or one positive roots and two or no negative roots**. The original equation has three sign changes, so there are three or one possible positive real roots. Then, substituting $-x$ for each x, $(-x)^5 - (-x)^3 + 8(-x)^2 - 8 = -x^5 + x^3 + 8x^2 - 8$ has two sign changes. There are two negative roots or none at all.

2 Count the number of possible positive and negative real roots in $8x^5 - 25x^4 - x^2 + 25 = 0$. The answer: **Two or no positive roots and one negative root**. It has two sign changes from positive to negative to positive so there are two or no positive roots. Substituting $-x$ for each x, $8(-x)^5 - 25(-x)^4 - (-x)^2 + 25 = -8x^5 - 25x^4 + x^2 + 25$ has one sign change and one negative root.

3 List all the possible rational roots of $2x^4 - 3x^3 - 54x + 81 = 0$. **The possible rational roots are** $\pm 81, \pm \frac{81}{2}, \pm 27, \pm \frac{27}{2}, \pm 9, \pm \frac{9}{2}, \pm 3, \pm \frac{3}{2}, \pm 1, \pm \frac{1}{2}$. The constant term in $2x^4 - 3x^3 - 54x + 81 = 0$ is 81. Its factors are $\pm 81, \pm 27, \pm 9, \pm 3, \pm 1$. The lead coefficient is 2 with factors $\pm 2, \pm 1$.

4 List all the possible rational roots of $8x^5 - 25x^3 - x^2 + 25 = 0$. **The possible rational roots are** $\pm 25, \pm \frac{25}{2}, \pm \frac{25}{4}, \pm \frac{25}{8}, \pm 5, \pm \frac{5}{2}, \pm \frac{5}{4}, \pm \frac{5}{8}, \pm 1, \pm \frac{1}{2}, \pm \frac{1}{4}, \pm \frac{1}{8}$. The constant term in $8x^5 - 25x^3 - x^2 + 25 = 0$ is 25, having factors $\pm 25, \pm 5, \pm 1$. The lead coefficient is 8 with factors $\pm 8, \pm 4, \pm 2, \pm 1$.

5 Check to see if -3 is a root of $x^4 - 10x^2 + 9 = 0$. **Yes**. Rewrite the equation with the coefficients showing in front of the variables: $x^4 - 10x^2 + 9 = 1(x^4) + 0(x^3) - 10(x^2) + 0(x) + 9$

Be sure to use 0 as a placeholder when a power of the variable is missing in the decreasing powers of the variable.

Using synthetic division:

```
-3| 1   0  -10   0   9
       -3    9   3  -9
    ─────────────────────
    1  -3   -1   3   0
```

Because the remainder is 0, the equation has a root of -3.

6 Check to see if 8 is a root of $x^5 - x^3 + 8x^2 - 8 = 0$. **No**. Use synthetic division:

```
8| 1  0  -1    8      0      -8
        8  64  504  4,096  32,768
   ──────────────────────────────
   1  8  63  512  4,096  32,760
```

You can see that 8 isn't a root, because the remainder isn't 0.

7 Check to see if $\frac{3}{2}$ is a root of $2x^4 - 3x^3 - 54x + 81 = 0$. **Yes**. Writing in the coefficients of the terms, you get $2x^4 - 3x^3 + 0(x^2) - 54x + 81 = 0$. Note that 3 is a factor of 81 and 2 is a factor of 2, so $\frac{3}{2}$ is a possible rational root:

```
3/2| 2  -3   0   -54   81
          3   0     0  -81
     ──────────────────────
     2   0   0   -54    0
```

The remainder is 0, so $\frac{3}{2}$ is a root (solution).

8 Check to see if –1 is a root of $8x^5 - 25x^3 - x^2 + 25 = 0$. **No.** Writing in the coefficients, $8x^5 + 0(x^4) - 25x^3 - x^2 + 0(x) + 25 = 0$, the synthetic division becomes

$$\underline{-1|}\ \ 8\ \ \ \ 0\ \ -25\ \ -1\ \ \ \ 0\ \ \ 25$$
$$\phantom{\underline{-1|}\ \ 8\ }\ -8\ \ \ \ 8\ \ \ \ 17\ -16\ \ 16$$
$$\overline{\phantom{\underline{-1|}}\ \ 8\ -8\ -17\ \ 16\ -16\ \ 41}$$

The remainder is 41, which isn't 0. So –1 isn't a root. Look at how you can check this by substituting –1 into the equation: $8(-1)^5 - 25(-1)^3 - (-1)^2 + 25 = -8 + 25 - 1 + 25 = 41 \neq 0$. Doing so again shows that –1 isn't a root.

9 Solve by factoring: $x^4 - 16 = 0$. The answer is ± 2. First factor the binomial as the difference and sum of the same two values; then factor the first of these factors the same way:

$$x^4 - 16 = 0$$
$$(x^2 - 4)(x^2 + 4) = 0$$
$$(x - 2)(x + 2)(x^2 + 4) = 0$$

$x - 2 = 0$, $x = 2$ or $x + 2 = 0$, $x = -2$ give the real solutions. So $x = \pm 2$. *Note:* $x^2 + 4 = 0$ has no real solutions.

10 Solve by factoring: $x^8 - x^2 = 0$. The answer is $x = 0, 0, \pm 1$.

$$x^8 - x^2 = 0$$
$$x^2(x^6 - 1) = 0$$

because x^2 is a common factor.

$$x^2(x^3 + 1)(x^3 - 1) = 0 \text{ because } x^6 = (x^3)^2$$
$$x^2(x + 1)(x^2 - x + 1)(x - 1)(x^2 + x + 1) = 0$$

so $x^2 = 0$, $x = 0, 0$; $x + 1 = 0$, $x = -1$; or $x - 1 = 0$, $x = 1$ give the real solutions. Therefore $x = 0, 0, \pm 1$.

11 Solve by factoring: $x^3 + 5x^2 - 16x - 80 = 0$. The answer is $x = \pm 4, -5$. First you get

$$x^3 + 5x^2 - 16x - 80 = 0$$
$$x^2(x + 5) - 16(x + 5) = 0$$

by grouping.

$$(x^2 - 16)(x + 5) = 0$$
$$(x - 4)(x + 4)(x + 5) = 0$$

so $x - 4 = 0$, $x = 4$; $x + 4 = 0$, $x = -4$; or $x + 5 = 0$, $x = -5$. Therefore $x = \pm 4, -5$.

12 Solve by factoring: $x^6 - 9x^4 - 16x^2 + 144 = 0$. The answer is $x = \pm 2, \pm 3$. First you get

$$x^6 - 9x^4 - 16x^2 + 144 = 0$$
$$x^4(x^2 - 9) - 16(x^2 - 9) = 0$$

by grouping.

Then each of these factors is the difference of perfect squares and can be factored into the difference and sum of the same two numbers:

$$(x^4 - 16)(x^2 - 9) = 0$$
$$(x^2 - 4)(x^2 + 4)(x^2 - 9) = 0$$
$$(x - 2)(x + 2)(x^2 + 4)(x - 3)(x + 3) = 0$$

So $x - 2 = 0$, $x = 2$; $x + 2 = 0$, $x = -2$; $x - 3 = 0$, $x = 3$; or $x + 3 = 0$, $x = -3$ give the real solutions. Therefore $x = \pm 2, \pm 3$.

13. Solve the equation: $x^4 - 13x^2 + 36 = 0$. The answer is $x = \pm 3, \pm 2$.

$x^4 - 13x^2 + 36 = 0$
$(x^2 - 9)(x^2 - 4) = 0$
so $x^2 - 9 = 0$, $x^2 = 9$, $x = \pm 3$ or $x^2 - 4 = 0$, $x^2 = 4$, $x = \pm 2$. Therefore $x = \pm 3, \pm 2$.

14. Solve the equation: $x^{10} - 31x^5 - 32 = 0$. The answer is $x = 2, -1$.

$x^{10} - 32x^5 - 32 = 0$
$(x^5 - 32)(x^5 + 1) = 0$
so $x^5 - 32 = 0$, $x^5 = 32$, $x = 2$ or $x^5 + 1 = 0$, $x^5 = -1$, $x = -1$.

15. Solve the equation: $y^{4/3} - 9y^{2/3} + 8 = 0$. The answer is $y = 1, 16\sqrt{2}$.

$y^{4/3} - 9y^{2/3} + 8 = 0$
$(y^{2/3} - 1)(y^{2/3} - 8) = 0$
so $y^{2/3} - 1 = 0$, $y^{2/3} = 1$, $\left(y^{2/3}\right)^{3/2} = (1)^{3/2}$, $y = 1$ or $y^{2/3} - 8 = 0$, $y^{2/3} = 8$, $\left(y^{2/3}\right)^{3/2} = 8^{3/2}$, $y = \left(2^3\right)^{3/2} = 2^{9/2} = 2^4 \times 2^{1/2} = 16\sqrt{2}$, therefore $y = 1, 16\sqrt{2}$.

16. Solve the equation: $z^{-2} + z^{-1} - 12 = 0$. The answer is $z = \frac{1}{3}, -\frac{1}{4}$.

$z^{-2} + z^{-1} - 12 = 0$
$(z^{-1} - 3)(z^{-1} + 4) = 0$
so $z^{-1} - 3 = 0$, $z^{-1} = 3$, $\frac{1}{z} = 3$, $1 = 3z$, $z = \frac{1}{3}$ or $z^{-1} + 4 = 0$, $z^{-1} = -4$, $\frac{1}{z} = -4$, $1 = -4z$, $z = -\frac{1}{4}$ therefore $z = \frac{1}{3}, -\frac{1}{4}$.

Chapter 15
Solving Radical and Absolute Value Equations

In This Chapter
▶ Squaring radicals once or twice
▶ Being absolutely sure of absolute value

Radical equations and absolute values are just what their names suggest. *Radical equations* contain one or more *radicals* (or square root symbols), while *absolute value equations* have an *absolute value operation* (two vertical bars that say to give the distance from 0). Although they definitely are two different types of equations, you may be wondering what they have in common. With them both, you change their form before solving. Changing a radical equation into a linear or quadratic equation (see Chapters 12 and 13), solving it, and considering the answers is easier than trying to develop a bunch of new rules and procedures. Going back to something familiar makes more sense.

The changing part of these two types of equations is what's different. I handle each type separately in this chapter and offer plenty of practice problems.

Solving Radical Equations by Squaring Once

If your radical equation has just one radical term, then you can solve it by isolating that radical term on one side of the equation and the other terms on the opposite side, and then squaring both sides.

Watch out for *extraneous roots*. These little devils crop up in several mathematical situations where you change from one type of equation to another, in order to solve the original equation. In this case, the squaring can introduce *extraneous* or false roots. Creating these false roots happens because the square of a number or its opposite gives you the same positive number. Squaring both sides to get these false answers may sound like a problem, but this procedure is still much easier than anything else. You really can't avoid the extraneous roots — just be aware that they can occur so you don't include them in your answer.

Actually, when trying to solve an equation with extraneous roots, you can encounter three possible outcomes:

✔ The most common thing to happen is that one of the two solutions works — the other solution turns out to be extraneous.
✔ Another possibility is that both of the solutions work.
✔ The third situation is that neither works.

Check out the following example to see how to handle an extraneous root in a radical equation.

Part III: Stirring Up Solutions

Q. Solve for x: $\sqrt{x+10} + x = 10$

A. $x = 6$

1. Move the x to the other side and square both sides of the equation.

   $\sqrt{x+10} = 10 - x$
   $(\sqrt{x+10})^2 = (10-x)^2$
   $x + 10 = 100 - 20x + x^2$

2. To solve this quadratic equation, set it equal to 0 and factor it.

   Two solutions appear.

   $0 = 90 - 21x + x^2$ or $x^2 - 21x + 90 = 0$
   $(x-15)(x-6) = 0$
   $x = 15$ or $x = 6$

3. Check for an extraneous solution.

   In this case, substituting the solutions, you see that the 6 works, but the 15 doesn't.

   $\sqrt{15+10} + 15 \stackrel{?}{=} 10$ $\sqrt{6+10} + 6 \stackrel{?}{=} 10$
   $\sqrt{25} + 15 \stackrel{?}{=} 10$ or $\sqrt{16} + 6 \stackrel{?}{=} 10$
   $5 + 15 \neq 10$ $4 + 6 = 10$

Q. What's the gestation period for an elephant?

A. **22 months**. Just imagine your mom being pregnant that long! Isn't that enough to appreciate her for what's she's done for you? (Oh gosh, go and give her a hug!)

---

1. Solve for x: $\sqrt{x-3} = 6$

Solve It

2. Solve for x: $\sqrt{x^2+9} = 5$

Solve It

3. Solve for x: $\sqrt{x+5} + x = 1$

Solve It

4. Solve for x: $\sqrt{x-3} + 9 = x$

Solve It

5. Solve for x: $\sqrt{x+7} - 7 = x$

Solve It

6. Solve for x: $\sqrt{x-1} = x - 1$

Solve It

Squaring Radical Equations Twice

Some radical equations have three or more terms — of which two or more are radical terms. In these cases, you can't isolate the radical term on one side. You have to square both sides, try to isolate the radical term on one side, and square both sides again. In general, squaring a *binomial* (the sum or difference of two terms) is easier if it has just one radical term, so I usually rewrite the equations, putting a radical term on each side before squaring.

Q. Solve for x: $\sqrt{5x+11} + \sqrt{x+3} = 2$

A. $x = -2$

1. Subtract $\sqrt{x+3}$ from each side.

   $\sqrt{5x+11} = 2 - \sqrt{x+3}$

2. Square both sides, simplify the terms, and get the radical term on one side and all the other terms on the opposite side.

   $\left(\sqrt{5x+11}\right)^2 = \left(2 - \sqrt{x+3}\right)^2$

   $5x + 11 = 4 - 4\sqrt{x+3} + x + 3$

   $5x + 11 = 7 + x - 4\sqrt{x+3}$

   $4x + 4 = -4\sqrt{x+3}$

3. As you can see, each of the terms is divisible by 4. Divide every term by 4, before squaring both sides, to keep the numbers smaller and more manageable.

   $(x+1)^2 = \left(-\sqrt{x+3}\right)^2$

   $x^2 + 2x + 1 = x + 3$

4. Set the quadratic equation equal to 0, factor it, and solve for the solutions to that equation.

   $x^2 + x - 2 = 0$

   $(x+2)(x-1) = 0$

   $x = -2$ or $x = 1$

5. Check the two solutions.

   You have two solutions, but do they both work? You need to test each one in the original equation to be sure it isn't an extraneous solution.

   $\sqrt{5(-2)+11} + \sqrt{(-2)+3} \stackrel{?}{=} 2$

   $\sqrt{1} + \sqrt{1} = 2$

   -2 is a solution.

   $\sqrt{5(1)+11} + \sqrt{(1)+3} \stackrel{?}{=} 2$

   $\sqrt{16} + \sqrt{4} \neq 2$

   The 1 is an extraneous root.

7. Solve for x: $\sqrt{x} + 3 = \sqrt{x + 27}$

Solve It

8. Solve for y: $\sqrt{3y + 1} - \sqrt{y + 4} = 1$

Solve It

9. Solve for z: $2\sqrt{z - 4} + \sqrt{z + 5} = 3$

Solve It

10. Solve for x: $3\sqrt{x + 1} - 2\sqrt{x - 4} = 5$

Solve It

Solving Absolute Value Equations

An absolute value equation contains an absolute value operation. Seems rather obvious, doesn't it? When solving an absolute value equation, remember what it has in common with radical equations. With both you change their form to solve them.

When solving an absolute value equation in the form $|ax + b| = c$, take the following steps:

1. **Rewrite the original equation as two separate equations.**

   $ax + b = c$ **or** $ax + b = -c$

2. **Solve these two equations separately, which results in two different answers.**

3. **Check the results in the original equation to ensure the answers work.**

   Generally, both answers work, but you need to check the results to be sure the original equation didn't have a nonsense statement (like having a positive equal to a negative) in it.

Q. Solve for x: $|4x + 5| = 3$

A. $x = -\frac{1}{2}$ **or** $x = -2$. Rewrite this as the two equations: $4x + 5 = 3$ or $4x + 5 = -3$. Solve $4x + 5 = 3$, which gives $x = -\frac{1}{2}$. Solve $4x + 5 = -3$, which gives $x = -2$. Check the first solution:

$$\left|4\left(-\frac{1}{2}\right) + 5\right| = |-2 + 5| = |3| = 3$$

Check the second solution:

$$|4(-2) + 5| = |-8 + 5| = |-3| = 3.$$

Q. How many donuts do you get when you buy a baker's dozen?

A. 13. Go figure. Who came up with this crazy definition? Consider yourself lucky. You get an extra one, but I'm curious. What are you doing eating 13 donuts by yourself?

11. Solve for x: $|x - 2| = 6$

Solve It

12. Solve for y: $|3y + 2| = 4$

Solve It

13. Solve for w: $|5w - 2| + 3 = 6$

Solve It

14. Solve for y: $3|4 - y| + 2 = 8$

Solve It

15. Solve for x: $|-4x| = 12$

Solve It

16. Solve for y: $|y + 3| + 6 = 2$

Solve It

Answers to Problems on Radical and Absolute Value Equations

This section provides the answers (in bold) to the practice problems in this chapter.

1 Solve for x: $\sqrt{x-3} = 6$. The answer is $x = \mathbf{39}$. First square both sides, then solve for x by adding 3 to each side:

$$\sqrt{x-3} = 6$$
$$\left(\sqrt{x-3}\right)^2 = 6^2$$
$$x - 3 = 36, \ x = 39$$

Then check:

$$\sqrt{39-3} \stackrel{?}{=} 6$$
$$\sqrt{36} = 6$$

2 Solve for x: $\sqrt{x^2+9} = 5$. The answer is $x = \pm \mathbf{4}$. First square both sides, then subtract 9 from each side. Find the square root of each side, and check to see if the answers work:

$$\sqrt{x^2+9} = 5$$
$$\left(\sqrt{x^2+9}\right)^2 = 5^2$$
$$x^2 + 9 = 25$$
$$x^2 = 16, \ x = \pm 4$$

Then check:

$$\sqrt{(4)^2 + 9} \stackrel{?}{=} 5, \ \sqrt{16+9} \stackrel{?}{=} 5, \ \sqrt{25} = 5$$
$$\sqrt{(-4)^2 + 9} \stackrel{?}{=} 5, \ \sqrt{16+9} \stackrel{?}{=} 5, \ \sqrt{25} = 5$$

This shows $x = \pm 4$.

3 Solve for x: $\sqrt{x+5} + x = 1$. The answer is $x = \mathbf{-1}$. First move x to the right side:

$$\sqrt{x+5} + x = 1$$
$$\sqrt{x+5} = 1 - x$$

Be sure to isolate the radical on one side of the equation or the other before squaring both sides of the equation.

Then square both sides. Set the equation equal to 0, so you can factor and solve for x:

$$\left(\sqrt{x+5}\right)^2 = (1-x)^2$$
$$x + 5 = 1 - 2x + x^2, \ 0 = x^2 - 3x - 4, \ (x-4)(x+1) = 0$$
$$x - 4 = 0, \ x = 4 \text{ or } x + 1 = 0, \ x = -1$$

Check: $x = 4$: $\sqrt{4+5} + 4 \stackrel{?}{=} 1, \ \sqrt{9} + 4 \stackrel{?}{=} 1, \ 3 + 4 \neq 1$. So 4 isn't a solution; it's extraneous.

Check: $x = -1$: $\sqrt{(-1)+5} + (-1) \stackrel{?}{=} 1, \ \sqrt{4} - 1 \stackrel{?}{=} 1, \ 2 - 1 = 1$. So $x = -1$ is the only solution.

Chapter 15: Solving Radical and Absolute Value Equations

4 Solve for x: $\sqrt{x-3}+9=x$. The answer is $x = 12$. First subtract 9 from each side and then square both sides. Set the quadratic equal to 0 to factor and solve for x:

$$\sqrt{x-3}+9=x$$
$$\sqrt{x-3}=x-9$$
$$\left(\sqrt{x-3}\right)^2=(x-9)^2$$
$$x-3=x^2-18x+81,\ 0=x^2-19x+84,\ (x-12)(x-7)=0$$

Using the multiplication property of zero (see Chapter 13), you have: $x - 12 = 0$, $x = 12$ or $x - 7 = 0$, $x = 7$

Check: $x = 12$: $\sqrt{12-3}+9 \stackrel{?}{=} 12$, $\sqrt{9}+9 \stackrel{?}{=} 12$, $3 + 9 = 12$ and $x = 7$: $\sqrt{7-3}+9 \stackrel{?}{=} 7$, $\sqrt{4}+9 \stackrel{?}{=} 7$, $2 + 9 \ne 7$

5 Solve for x: $\sqrt{x+7}-7=x$. The answer is both $x = -7$ and $x = -6$.

First add 7 to each side. Then square both sides, set the equation equal to 0, and solve for x:

$$\sqrt{x+7}-7=x$$
$$\sqrt{x+7}=x+7$$
$$\left(\sqrt{x+7}\right)^2=(x+7)^2$$
$$x+7=x^2+14x+49,\ 0=x^2+13x+42,\ (x+7)(x+6)=0$$

Using the multiplication property of zero (see Chapter 13), you have $x + 7 = 0$, $x = -7$ or $x + 6 = 0$, $x = -6$.

Check: $x = -7$: $\sqrt{(-7)+7}-7 \stackrel{?}{=} -7$, $\sqrt{0}-7 \stackrel{?}{=} -7$, $0 - 7 = -7$ and $x = -6$: $\sqrt{(-6)+7}-7 \stackrel{?}{=} -6$, $\sqrt{1}-7 \stackrel{?}{=} -6$, $1 - 7 = -6$

6 Solve for x: $\sqrt{x-1}=x-1$. The answer is both $x = 1$ and $x = 2$. First square both sides of the equation; then set it equal to 0 and factor:

$$\left(\sqrt{x-1}\right)^2=(x-1)^2$$
$$x-1=x^2-2x+1$$
$$0=x^2-3x+2$$
$$0=(x-1)(x-2)$$
$$x=1\ \text{or}\ x=2$$

Using the multiplication property of zero (see Chapter 13), you have $x - 1 = 0$, $x = 1$ or $x - 2 = 0$, $x = 2$.

Check: $x = 1$, $\sqrt{1-1} \stackrel{?}{=} 1-1$, $0 = 0$ and $x = 2$, $\sqrt{2-1} \stackrel{?}{=} 2-1$, $1 = 1$

7 Solve for x: $\sqrt{x}+3=\sqrt{x+27}$. The answer is $x = 9$. First square both sides of the equation. Then keep the radical term on the left and subtract x and 9 from each side. Before squaring both sides again, divide by 6:

$$\sqrt{x}+3=\sqrt{x+27}$$
$$\left(\sqrt{x}+3\right)^2=\left(\sqrt{x+27}\right)^2$$
$$\left(\sqrt{x}\right)^2+6\sqrt{x}+9=x+27$$
$$x+6\sqrt{x}+9=x+27$$
$$6\sqrt{x}=18$$
$$\sqrt{x}=3$$

Square both sides of the new equation: $\left(\sqrt{x}\right)^2=3^2$, $x = 9$

Check: $x = 9$: $\sqrt{9}+3 \stackrel{?}{=} \sqrt{9+27}$, $3+3=6=\sqrt{36}$

8 Solve for y: $\sqrt{3y+1} - \sqrt{y+4} = 1$. The answer is $y = 5$. First move $\sqrt{y+4}$ to the other side:

$$\sqrt{3y+1} - \sqrt{y+4} = 1$$
$$\sqrt{3y+1} = \sqrt{y+4} + 1$$

Then square both sides of the equation, simplify, and isolate the radical term on the right. Both sides are divisible by 2, so divide:

$$\left(\sqrt{3y+1}\right)^2 = \left(\sqrt{y+4} + 1\right)^2$$
$$3y + 1 = \left(\sqrt{y+4}\right)^2 + 2\sqrt{y+4} + 1$$
$$3y + 1 = y + 4 + 2\sqrt{y+4} + 1$$
$$3y + 1 = y + 5 + 2\sqrt{y+4}$$
$$2y - 4 = 2\sqrt{y+4}$$
$$y - 2 = \sqrt{y+4}$$

Then square both sides of the new equation:

$$(y-2)^2 = \left(\sqrt{y+4}\right)^2$$
$$y^2 - 4y + 4 = y + 4, \ y^2 - 5y = 0$$
$$y(y-5) = 0$$

Using the multiplication property of zero (see Chapter 13), you have $y = 0$ or $y - 5 = 0$, $y = 5$.

Check: $y = 0$: $\sqrt{3(0)+1} - \sqrt{0+4} = \sqrt{1} - \sqrt{4} = 1 - 2 = -1 \neq 1$ and $y = 5$: $\sqrt{3(5)+1} - \sqrt{5+4} = \sqrt{16} - \sqrt{9} = 4 - 3 = 1$.

9 Solve for z: $2\sqrt{z-4} + \sqrt{z+5} = 3$. The answer is $z = 4$. First move the second radical term to the right. Then square both sides, simplify the terms, and move everything except the radical term to the left. Divide each side by 3:

$$2\sqrt{z-4} + \sqrt{z+5} = 3$$
$$2\sqrt{z-4} = 3 - \sqrt{z+5}$$
$$\left(2\sqrt{z-4}\right)^2 = \left(3 - \sqrt{z+5}\right)^2$$
$$4(z-4) = 9 - 6\sqrt{z+5} + (z+5)$$
$$4z - 16 = z + 14 - 6\sqrt{z+5}$$
$$3z - 30 = -6\sqrt{z+5}$$
$$z - 10 = -2\sqrt{z+5}$$

Then square both sides. Set the equation equal to 0 and factor:

$$(z-10)^2 = \left(-2\sqrt{z+5}\right)^2$$
$$z^2 - 20z + 100 = 4(z+5)$$
$$z^2 - 20z + 100 = 4z + 20$$
$$z^2 - 24z + 80 = 0, \ (z-20)(z-4) = 0$$

Using the multiplication property of 0, $z = 20$ or $z = 4$.

Check: $z = 20$: $2\sqrt{20-4} + \sqrt{20+5} \stackrel{?}{=} 3$, $2\sqrt{16} + \sqrt{25} = 2(4) + 5 \neq 3$ and $z = 4$: $2\sqrt{4-4} + \sqrt{4+5} \stackrel{?}{=} 3$, $2\sqrt{0} + \sqrt{9} = 0 + 3 = 3$

Chapter 15: Solving Radical and Absolute Value Equations

10. Solve for x: $3\sqrt{x+1} - 2\sqrt{x-4} = 5$. The answer is $x = 8$. First move a radical term to the right, square both sides, simplify, and, finally, isolate the radical term on the right. You can then divide each side by 5:

$$3\sqrt{x+1} - 2\sqrt{x-4} = 5$$
$$3\sqrt{x+1} = 2\sqrt{x-4} + 5$$
$$\left(3\sqrt{x+1}\right)^2 = \left(2\sqrt{x-4} + 5\right)^2$$
$$9(x+1) = 4(x-4) + 20\sqrt{x-4} + 25$$
$$9x + 9 = 4x - 16 + 20\sqrt{x-4} + 25$$
$$9x + 9 = 4x + 9 + 20\sqrt{x-4}$$
$$5x = 20\sqrt{x-4}$$
$$x = 4\sqrt{x-4}$$

Square both sides again, and set the equation equal to 0 and factor:

$$x^2 = \left(4\sqrt{x-4}\right)^2$$
$$x^2 = 16(x-4)$$
$$x^2 = 16x - 64, \; x^2 - 16x + 64 = 0, \; (x-8)^2 = 0$$

Using the multiplication property of zero (see Chapter 13), you have $x - 8 = 0$, $x = 8$.

Check:

$x = 8$: $3\sqrt{8+1} - 2\sqrt{8-4} \stackrel{?}{=} 5$
$3\sqrt{9} - 2\sqrt{4} \stackrel{?}{=} 5$, $3(3) - 2(2) = 9 - 4 = 5$

11. Solve for x: $|x - 2| = 6$. The answer is $x = 8$ or $x = -4$. First remove the absolute value symbol by setting what's inside equal to both positive and negative 6. Then solve the two linear equations that can be formed:

$|x - 2| = 6$
$x - 2 = \pm 6$
$x = 2 \pm 6; \; x = 2 + 6 = 8$ or $x = 2 - 6 = -4$

Check: $x = 8$: $|8 - 2| = |6| = 6$ and $x = -4$: $|(-4) - 2| = |-6| = 6$

12. Solve for y: $|3y + 2| = 4$. The answer is $y = \frac{2}{3}$ or $y = -2$. First remove the absolute value symbol by setting what's inside equal to both positive and negative 4. Then solve the two linear equations that can be formed:

$|3y + 2| = 4$
$3y + 2 = \pm 4, \; 3y = -2 \pm 4$
$3y = -2 + 4$ gives $3y = 2, \; y = \frac{2}{3}$
$3y = -2 - 4$ gives $3y = -6, \; y = -2$

Check: $y = \frac{2}{3}$: $\left|3\left(\frac{2}{3}\right) + 2\right| = |2 + 2| = |4| = 4$ and $y = -2$: $|3(-2) + 2| = |-6 + 2| = |-4| = 4$

13. Solve for w: $|5w - 2| + 3 = 6$. The answer is $w = 1$ or $w = -\frac{1}{5}$. First subtract 3 from each side. Then remove the absolute value symbol by setting what's inside equal to both positive and negative 3:

$|5w - 2| + 3 = 6$
$|5w - 2| = 6 - 3 = 3$
$5w - 2 = \pm 3, \; 5w = 2 \pm 3$

The two linear equations that are formed give you two different answers:

$5w = 2 + 3$ gives $5w = 5$, $w = 1$

$5w = 2 - 3$ given $5w = -1$

$w = -\frac{1}{5}$

Check: $w = 1$: $|5(1) - 2| + 3 \stackrel{?}{=} 6$, $|3| + 3 \stackrel{?}{=} 6$, $3 + 3 = 6$ and $w = -\frac{1}{5}$: $\left|5\left(-\frac{1}{5}\right) - 2\right| + 3 \stackrel{?}{=} 6$, $|-3| + 3 \stackrel{?}{=} 6$, $3 + 3 = 6$

14 Solve for y: $3|4 - y| + 2 = 8$. The answer is **$y = 6$ or $y = 2$**. First subtract 2 from each side. Then divide each side by 3:

$3|4 - y| + 2 = 8$

$3|4 - y| = 6$

$|4 - y| = 2$

Then rewrite without the absolute value symbol by setting the expression inside the absolute value equal to positive or negative 2: $4 - y = \pm 2$; $4 \pm 2 = y$. The resulting linear equations are then solved: $y = 4 + 2 = 6$ or $y = 4 - 2 = 2$.

Check: $y = 6$: $3|4 - 6| + 2 = 3|-2| + 2 = 3(2) + 2 = 8$ and $y = 2$: $3|4 - 2| + 2 = 3|2| + 2 = 3(2) + 2 = 8$

15 Solve for x: $|-4x| = 12$. The answer is **$x = 3$ or $x = -3$**. First rewrite the equation without the absolute value symbol:

$|-4x| = 12$

$-4x = \pm 12$

$-4x = 12$ gives $x = -\frac{12}{4} = -3$

$-4x = -12$ gives $x = -\frac{-12}{4} = 3$

Check: $x = -3$: $|-4(-3)| = |12| = 12$ and $x = 3$: $|-4(3)| = |-12| = 12$

So $x = \pm 3$.

16 Solve for y: $|y + 3| + 6 = 2$. **This equation doesn't have an answer.** Here's why:

$|y + 3| + 6 = 2$

$|y + 3| = -4$

$y + 3 = \pm 4$, $y = -3 \pm 4$, $y = 1$ or $y = -7$

Check: $y = 1$: $|(1) + 3| + 6 = |4| + 6 = 4 + 6 = 10 \neq 2$ and $y = -7$: $|(-7) + 3| + 6 = |-4| + 6 = 4 + 6 = 10 \neq 2$

Note that $|y + 3| = -4$ is impossible, because $|y + 3| \geq 0$ can never equal a negative number.

Chapter 16
Working with Inequalities

In This Chapter
- Playing by the rules for inequalities
- Solving linear and quadratic inequalities
- Dealing with absolute value inequalities
- Working with sectioned inequalities

An *inequality* is a mathematical statement that says that some number or variable is bigger or smaller than another number or variable. Sometimes the inequality includes an equal sign to show that you want something *bigger than or equal to,* or *smaller than or equal to.*

The good news about solving inequalities is that *nearly* all the rules are the same as solving an equation with one big difference. The difference comes in when you're multiplying or dividing with *negative* numbers. If you pay attention to what you're doing, you shouldn't have a problem.

This chapter covers everything from your run-of-the-mill inequalities and linear equalities to the more difficult quadratic, absolute value, and section inequalities. Take a deep breath. I offer you plenty of practice problems so you can work out any kinks.

Using the Rules to Work on Inequality Statements

Working with inequalities really isn't that difficult if you can keep a few rules in mind. The following rules deal with inequalities (assume that c is some number):

If $a > b$, then adding c to each side gives you $a + c > b + c$.

If $a > b$, then subtracting c from each side gives you $a - c > b - c$.

If $a > b$, then multiplying each side by a *positive* c gives you $ac > bc$.

If $a > b$, then multiplying each side by a *negative* c gives you $ac < bc$.

If $a > b$, then dividing each side by a *positive* c gives you $\frac{a}{c} > \frac{b}{c}$.

If $a > b$, then dividing each side by a *negative* c gives you $\frac{a}{c} < \frac{b}{c}$.

If $a > b$, then reversing the inequality reverses the terms $b < a$.

Q. Use –4 < 5 and perform the following operations: Add 3 to each side, multiply each side by –2, divide each side by 6, and then reverse the inequality.

A. $-\frac{8}{3} < \frac{1}{3}$

$$-4 < 5$$
$$-4 + 3 < 5 + 3 \text{ or } -1 < 8$$
$$-1(-2) > 8(-2) \text{ or } 2 > -16$$
$$\frac{2}{6} > -\frac{16}{6} \text{ or } \frac{1}{3} > -\frac{8}{3} \text{ or } -\frac{8}{3} < \frac{1}{3}$$

The inequality sign changes only when I multiply by the –2. Even though I perform all these operations, the statement is always true, because the beginning statement was true and the rules were followed.

1. Starting with 5 > 2, add 4 to each side and simplify the result.

Solve It

2. Starting with 5 ≥ 1, multiply each side by 4 and simplify the result.

Solve It

3. Starting with –5 ≤ –1, multiply each side by –3 and simplify the result.

Solve It

4. Starting with –6 < 10, divide each side by –2, multiply each side by –1, and simplify your result.

Solve It

Solving Linear Inequalities

Solving linear equalities involves pretty much the same process as solving linear equations — move all the variable terms to one side and the number to the other side. Then multiply or divide to solve for the variable. The tricky part is when you multiply or divide by a negative number. Because it's a rule used in special cases, people tend to forget it. **Remember:** If you multiply both sides of $-x < -3$ by –1, the inequality becomes $x > 3$; you have to reverse the sense.

Another type of linear inequality has the linear expression sandwiched between two numbers, like $-1 < 8 - x \leq 4$. The main rule here is that whatever you do to one *section,* you do to all the others. For more on this, go to "Solving Inequalities by Sections," later in this chapter.

Q. Solve for x: $5 - 3x \geq 14 + 6x$

A. $x \leq -1$. Subtract $6x$ from each side and subtract 5 from each side, then divide by -9:

$$5 - 3x \geq 14 + 6x$$
$$\underline{-5 - 6x \quad -5 - 6x}$$
$$-9x \geq 9$$
$$\frac{-9x}{-9} \leq \frac{9}{-9}$$
$$x \leq -1$$

But, look at this same problem where I try to avoid dividing by a negative number. I add $3x$ to each side and subtract 14 from each side. Then I divide by 9. This is the same answer, if I reverse the inequality and the numbers.

$$5 - 3x \geq 14 + 6x$$
$$\underline{-14 + 3x \quad -14 + 3x}$$
$$-9 \geq 9x$$
$$\frac{-9}{9} \geq \frac{9x}{9}$$
$$-1 \geq x$$

Q. Solve for x: $-1 < 8 - x \leq 4$

A. $4 \leq x < 9$. Subtract 8 from each section, and then divide each by -1. Write the final answer reading from the smaller number to the larger.

$$-1 < 8 - x \leq 4$$
$$\underline{-8 \quad -8 \quad -8}$$
$$-9 < -x \leq -4$$
$$\frac{-9}{-1} > \frac{-x}{-1} \geq \frac{-4}{-1}$$
$$9 > x \geq 4$$

5. Solve for x: $2x + 1 > 3$

Solve It

6. Solve for y: $4 - 5y \leq 19$

Solve It

7. Solve for x: $3(x + 2) > 4x + 5$

Solve It

8. Solve for x: $2x(3x + 7) < 6(x^2 - 1)$

Solve It

9. Solve for x: $-5 < 2x + 3 \leq 7$

Solve It

10. Solve for x: $3 \leq 7 - 2x < 11$

Solve It

Solving Quadratic Inequalities

When an inequality involves a quadratic expression, you have to resort to a completely different type of process to solve it. The problem arises, because you get a positive result when multiplying two positives together and, also, when you multiply two negatives together. So, how do you know which it was if you just have a positive result? Rather than go into a long, involved process with several equations, use a number line to solve these problems.

If you need to see for yourself, you can use a number line and place some + and – signs to indicate what's happening to the factors. (See Chapter 1 for some background info on the number line and where numbers fall on it.)

Q. Solve the inequality: $x^2 - x > 12$

A. $x < -3$ or $x > 4$

1. **Subtract 12 from each side to have it greater than 0.**

2. **Factor the quadratic and determine the *zeros*.**

   You want to find out what values of x make the different factors equal to 0, so you can separate intervals on the number line into positive and negative portions for each factor.

   $$x^2 - x - 12 > 0$$
   $$(x-4)(x+3) > 0$$

   The first factor is equal to 0 when x is equal to 4. The second factor is 0 when x is -3.

3. **Mark these two numbers on the number line.**

   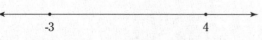

4. **Look to the left of the –3, between the –3 and 4, and then to the right of the 4** and indicate, on the number line, what the signs of the factors are in those intervals.

   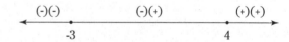

   When you replace the x with any number smaller than –3, the factor $(x - 4)$ is negative, and the factor $(x + 3)$ is also negative. When you choose a number between –3 and 4, say 0, for instance, the factor $(x - 4)$ is negative, and the factor $(x + 3)$ is positive. When you replace the x with any number bigger than 4, the factor $(x - 4)$ is positive, and the factor $(x + 3)$ is also positive.

   Recall that, when you multiply signed numbers with the same signs, the product is positive, and when the signs are different, the product is negative. The original problem asks for when the product is positive, and, according to the signs on the line, this is when $x < -3$ or $x > 4$.

Chapter 16: Working with Inequalities — 193

11. Solve for x: $(x+7)(x-1) > 0$

Solve It

12. Solve for x: $x^2 - x \leq 20$

Solve It

13. Solve for x: $3x^2 \geq 9x$

Solve It

14. Solve for x: $x^2 - 2 \geq 0$

Solve It

Solving Other Inequalities

Two other types of inequality problems use the same process as in the quadratic inequalities. (See "Solving Quadratic Inequalities" earlier in this chapter for more info.) You can use the number line process for higher degree expressions (solving polynomial inequalities) and for fractions (solving rational inequalities) — where a variable ends up in the *denominator* (bottom of fraction).

Q. Solve for x:
$(x+3)(x+2)(x-1)^2(x-8) \leq 0$

A. $x \leq -3$ or $-2 \leq x \leq 8$. Set each factor equal to 0 to determine where the factors might change from positive to negative or vice versa. Place the factors on a number line, test each factor in each interval on the number line, and determine the sign of the expression in that interval. Use that information to solve the problem.

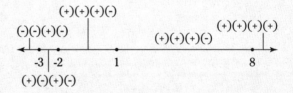

The solution consists of all the numbers that make the expression either negative or equal to 0. Notice that there was no change in the sign of the product when x went from smaller than 1 to larger than 1. That's because that factor was squared, making that factor either positive or 0 all the time.

Q. Solve for x: $\frac{x+2}{x-3} \geq 0$

A. $x \leq -2$ or $x > 3$. Solving an inequality with a fraction works pretty much the same way. Find all the zeros of the factors, put them on a number line, and check for the sign of the result. The only thing you need to be careful of is not including some number that would result in a 0 factor in the denominator. The sign can change from one side to the other of that number — you just can't use it in your answer. See this number line for clarification.

Remember the rule for dividing a signed number is the same as it is in multiplication. According to the sign line, the fraction is positive for numbers smaller than –2 or for numbers greater than +3. Also, you get a 0 when $x = -2$.

15. Solve for x: $x(x-1)(x+2) \geq 0$

Solve It

16. Solve for x: $x^4 - 26x^2 + 25 < 0$

Solve It

17. Solve for x: $x^3 - 4x^2 + 4x - 16 \leq 0$

Solve It

18. Solve for x: $\frac{5+x}{x} > 0$

Solve It

19. Solve for x: $\dfrac{x^2-1}{x+3} \le 0$

Solve It

20. Solve for x: $\dfrac{x+4}{2} + \dfrac{6}{x} \ge 6$

Hint: Move everything to the left and find a common denominator. Then add/subtract the terms to form a single fraction.

Solve It

Solving Absolute Value Inequalities

Algebra has absolute value equations and inequalities. Put them together and you get *absolute value inequalities.* To solve these problems, you need to rewrite the form to change them into simpler inequality problems — types you already know how to solve. Solve the new problem or problems for the solution to the original problem.

You need to keep two different situations in mind. (Always assume that the *c* is a positive number or 0.)

- If you have $|ax+b| > c$, change it to $ax + b > c$ or $ax + b < -c$ and solve the two inequalities.
- If you have $|ax+b| < c$, change it to $-c < ax + b < c$ and solve the one compound inequality.

Q. Solve for x: $|9x-5| > 4$

A. $x < \frac{1}{9}$ or $x > 1$. Change the inequality to $9x - 5 > 4$ or $9x - 5 < -4$. Solving the two inequalities, you get the two solutions $9x > 9, x > 1$ or $9x < 1, x < \frac{1}{9}$. So numbers bigger than 1 or smaller than $\frac{1}{9}$ satisfy the original statement.

Q. Solve for x: $|9x-5| < 4$

A. $\frac{1}{9} < x < 1$. This is like the first example, except that the sense has been turned around. Rewrite it as $-4 < 9x - 5 < 4$. Solving it, the solution is $1 < 9x < 9$, $\frac{1}{9} < x < 1$. The inequality statement says that all the numbers between $\frac{1}{9}$ and 1 work. Notice that the interval in this answer is what was left out in writing the solution to the problem when the inequality was reversed — that and the exact numbers $\frac{1}{9}$ and 1.

21. Solve for x: $|4x - 3| < 5$

Solve It

22. Solve for x: $|2x + 1| \geq 7$

Solve It

23. Solve for x: $|x - 3| + 6 \leq 8$

Solve It

24. Solve for x: $5|7x - 4| + 1 > 6$

Solve It

Solving Inequalities by Sections

A *compound inequality* — one with more than two *sections* (intervals or expressions sandwiched between inequalities) to it — could have variables in more than one section where adding and subtracting can't isolate the variable in one place. When this happens, you have to break up the inequality into solvable sections and write the answer in terms of what the different solutions share.

Q. Solve for x: $1 \leq 4x - 3 < 3x + 7$

A. $1 \leq x < 10$. Break it up into the two separate problems: $1 \leq 4x - 3$ and $4x - 3 < 3x + 7$. Solve the first inequality to get $4 \leq 4x$, $1 \leq x$. Solve the second inequality to get $x - 3 < 7$, $x < 10$. Putting these two answers together, you get that x must be some number bigger than or equal to 1 while, at the same time, smaller than 10, which leaves several numbers to use in the solution.

Q. Jill weighs less than Jack who weighs as much as or less than the giant. If they each lose at least ten pounds, then how many oranges equal a banana?

A. **Six**. What? You didn't get that? Good. I would have worried about you if you had. Sometimes you just have to back up and see if a question makes sense or not. Of course, all of my problems and questions are perfectly clear. I just threw in this one for contrast.

25. Solve for x: $6 \leq 5x + 1 < 2x + 10$

Solve It

26. Solve for x: $-6 \leq 4x - 3 < 5x + 1$

Solve It

Answers to Problems on Working with Inequalities

This section provides the answers (in bold) to the practice problems in this chapter.

1 Starting with 5 > 2, add 4 to each side and simplify the result. The answer is **9 > 6**.

$$5 > 2$$
$$5 + 4 > 2 + 4$$
$$9 > 6$$

2 Starting with $5 \geq 1$, multiply each side by 4 and simplify the result. The answer is $\mathbf{20 \geq 4}$, because 4 > 0.

3 Starting with $-5 \leq -1$, multiply each side by -3 and simplify the result. The answer is $\mathbf{15 \geq 3}$.

$$-5 \leq -1$$
$$(-3)(-5) \geq (-1)(-3)$$

because $-3 < 0$. Change the direction of $\leq$ to get $15 \geq 3$.

4 Starting with $-6 < 10$, divide each side by -2, multiply each side by -1, and simplify your result. The answer is $\mathbf{-3 < 5}$.

$$-6 < 10$$
$$\frac{-6}{-2} > \frac{10}{-2}$$

because $-2 < 0$.

$$3 > -5$$
$$(-1)3 < (-1)(-5)$$

because $-1 < 0$, which gives you $-3 < 5$.

5 Solve for x: $2x + 1 > 3$. The answer is $\mathbf{x > 1}$.

$$2x + 1 > 3$$
$$\underline{-1 -1}$$
$$2x > 2$$
$$\frac{2x}{2} > \frac{2}{2}$$
$$x > 1$$

6 Solve for y: $4 - 5y \leq 19$. The answer is $\mathbf{y \geq -3}$. First subtract 4 from both sides, and then divide by -5 to get the answer:

$$4 - 5y \leq 19$$
$$\underline{-4 -4}$$
$$-5y \leq 15$$
$$\frac{-5y}{-5} \geq \frac{15}{-5}$$
$$y \geq -3$$

Chapter 16: Working with Inequalities

7 Solve for x: $3(x + 2) > 4x + 5$. The answer is **$1 > x$ (or $x < 1$)**.

$$3(x+2) > 4x+5$$
$$3x+6 > 4x+5$$
$$\underline{-3x-5 \quad -3x-5}$$
$$1 > x \quad \text{or} \quad x < 1$$

8 Solve for x: $2x(3x + 7) < 6(x^2 - 1)$. The answer is **$x < -\frac{3}{7}$**.

$$2x(3x+7) < 6(x^2-1)$$
$$6x^2+14x < 6x^2-6$$
$$\underline{-6x^2 \quad\quad -6x^2}$$
$$14x < -6$$
$$\frac{14x}{14} < \frac{-6}{14}$$
$$x < -\frac{3}{7}$$

9 Solve for x: $-5 < 2x + 3 \le 7$. The answer is **$-4 < x \le 2$**.

$$-5 < 2x+3 \le 7$$
$$\underline{-3 \quad\quad -3 \quad -3}$$
$$-8 < 2x \quad\quad \le 4$$
$$\frac{-8}{2} < \frac{2x}{2} \quad \le \frac{4}{2}$$
$$-4 < x \quad\quad \le 2$$

10 Solve for x: $3 \le 7 - 2x < 11$. The answer is **$2 \ge x > -2$ (or $-2 < x \le 2$)**. First subtract 7, and then divide by -2 to get the answer:

$$3 \le 7-2x < 11$$
$$\underline{-7 \quad -7 \quad\quad -7}$$
$$-4 \le \quad -2x < 4$$
$$\frac{-4}{-2} \ge \frac{-2x}{-2} > \frac{4}{-2}$$
$$2 \ge x > -2 \quad \text{or} \quad -2 < x \le 2$$

11 Solve for x: $(x + 7)(x - 1) > 0$. The answer is **$x > 1$ or $x < -7$**.

TIP

Draw a number line to help you solve problems 11 through 20. For example, $x + 7 = 0$ when $x = -7$ and $x - 1 = 0$ when $x = 1$ as shown on the following number line.

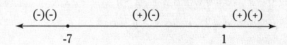

If $x > 1$, then both factors are positive. If $-7 < x < 1$, then $x + 7$ is positive, but $x - 1$ is negative, making the product negative. If $x < -7$, then both factors are negative, and the product is positive. So $x > 1$ or $x < -7$.

12 Solve for x: $x^2 - x \le 20$. The answer is **$-4 \le x \le 5$**. First subtract 20 from both sides, then factor the quadratic and set the factors equal to 0 to find the values where the factors change signs:

$$x^2 - x \leq 20$$
$$-20 \quad -20$$
$$\overline{x^2 - x - 20 \leq 0}$$
$$(x-5)(x+4) \leq 0$$

$x - 5 = 0$ when $x = 5$ and $x + 4 = 0$ when $x = -4$ as shown on the following number line.

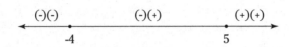

If $x > 5$, then $x - 5 > 0$ and $x + 4 > 0$, making the product positive. If $-4 < x < 5$, then $x - 5 < 0$ and $x + 4 > 0$, resulting in a product that's negative. If $x < -4$, then both factors are negative, and the product is positive. So the product is negative only if $-4 < x < 5$. But the solutions of $(x - 5)(x + 4) = 0$ are $x = 5$, $x = -4$. Including these two values, the solution is then $-4 \leq x \leq 5$.

13 Solve for x: $3x^2 \geq 9x$. **The answer is $x \geq 3$ or $x \leq 0$.** First subtract $9x$ from each side. Then factor the quadratic and set the factors equal to 0:

$$3x^2 \geq 9x$$
$$3x^2 - 9x \geq 0$$
$$3x(x - 3) \geq 0$$

$3x(x - 3) = 0$ when $x = 0$, $x = 3$ as shown on the following number line.

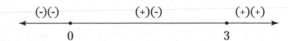

If $x > 3$, both factors are positive, and $3x(x - 3) > 0$. If $0 < x < 3$, then $3x > 0$ and $x - 3 < 0$, making the product negative. If $x < 0$, both factors are negative, and $3x(x - 3) > 0$. When $x = 0$ or $x = 3$, $3x(x - 3) = 0$. Include those in the solution to get $x \geq 3$ or $x \leq 0$.

14 Solve for x: $x^2 - 2 \geq 0$. **The answer is $x \geq \sqrt{2}$ or $x \leq -\sqrt{2}$.** First set the expression equal to 0 to help find the factors:

$$x^2 - 2 \geq 0$$
$$x^2 - 2 = 0, \; x^2 = 2, \; x = \pm\sqrt{2}$$

Factor to get $x^2 - 2 = (x - \sqrt{2})(x + \sqrt{2})$, so the inequality becomes $(x - \sqrt{2})(x + \sqrt{2}) \geq 0$ as shown on the following number line.

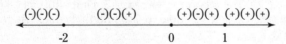

From the figure, $x^2 - 2 > 0$ when $x > \sqrt{2}$ or $x < -\sqrt{2}$.
When $x = \pm\sqrt{2}$ then $x^2 - 2 = 0$. So $x \geq \sqrt{2}$ or $x \leq -\sqrt{2}$.

15 Solve for x: $x(x - 1)(x + 2) \geq 0$. **The answer is $x \geq 1$ or $-2 \leq x \leq 0$.** First set the factored expression equal to 0 to find where the factors change signs. Put these values on the number line:

$$x(x - 1)(x + 2) \geq 0$$
$$x(x - 1)(x + 2) = 0 \text{ when } x = 0, 1, -2$$

```
       (-)(-)(-)    (-)(-)(+)    (+)(-)(+)  (+)(+)(+)
    ←─────●──────────●───────────●─────────●─────────→
          -2         0           1
```

Assign signs to each of the four regions to get the answer: $x \geq 1$ or $-2 \leq x \leq 0$.

16 Solve for x: $x^4 - 26x^2 + 25 < 0$. The answer is $1 < x < 5$ or $-5 < x < -1$. First factor the expression and set it equal to 0. Solve the equations that can be formed to find out where the factors change sign. Put these values on the number line:

$$x^4 - 26x^2 + 25 < 0$$
$$(x^2 - 25)(x^2 - 1) < 0$$
$$(x-5)(x+5)(x-1)(x+1) < 0$$
$$(x-5)(x+5)(x-1)(x+1) = 0 \text{ when } x = \pm 5 \text{ or } \pm 1$$

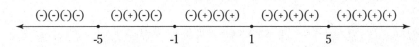

Assign signs to each of the five intervals to get the answer: $-5 < x < -1$ or $1 < x < 5$.

17 Solve for x: $x^3 - 4x^2 + 4x - 16 \leq 0$. The answer is $x \leq 4$. First factor the expression by grouping. Then set the factors equal to 0.

$$x^3 - 4x^2 + 4x - 16 \leq 0$$
$$x^2(x-4) + 4(x-4) \leq 0$$
$$(x^2 + 4)(x-4) \leq 0$$
$$(x^2 + 4)(x-4) = 0$$

only if $x = 4$. Put this number on the number line.

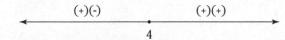

Assign signs to each interval to get the answer $x \leq 4$.

18 Solve for x: $\frac{5+x}{x} > 0$. The answer is $x > 0$ or $x < -5$.

$$\frac{5+x}{x} > 0$$
$$\frac{5+x}{x} = 0$$

when $5 + x = 0$, $x = -5$.

$\frac{5+x}{x}$ is undefined when the denominator $x = 0$, so place $x = 0$ and $x = -5$ on the number line.

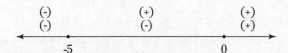

Assign signs to each of the three intervals to get the answer: $x > 0$ or $x < -5$.

19 Solve for x: $\frac{x^2-1}{x+3} \leq 0$. The answer is $-1 \leq x \leq 1$ or $x < -3$.

$$\frac{x^2-1}{x+3} \leq 0$$
$$\frac{(x-1)(x+1)}{x+3} \leq 0$$
$$\frac{(x-1)(x+1)}{x+3} = 0$$

when $x = 1$ or $x = -1$ and undefined when $x = -3$.

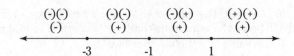

Assign signs to each of the four intervals. So $\frac{(x-1)(x+1)}{x+3} < 0$ when $-1 < x < 1$ or $x < -3$. But $\frac{(x-1)(x+1)}{x+3} = 0$ only when $x = 1$ or $x = -1$, not at $x = -3$. So the way to write these answers all together with inequality symbols is: $-1 \leq x \leq 1$ or $x < -3$

20 Solve for x: $\frac{x+4}{2} + \frac{6}{x} \geq 6$. The answer is $\boldsymbol{x \geq 6}$ **or** $\boldsymbol{0 < x \leq 2}$.

$$\frac{x+4}{2} + \frac{6}{x} \geq 6$$
$$\underline{-6-6}$$
$$\frac{x+4}{2} + \frac{6}{x} - 6 \geq 0$$
$$\frac{x+4}{2}\left(\frac{x}{x}\right) + \frac{6}{x}\left(\frac{2}{2}\right) - 6\left(\frac{2x}{2x}\right) \geq 0$$

where the common denominator is $2x$.

$$\frac{(x+4)x + 12 - 12x}{2x} \geq 0$$
$$\frac{x^2 + 4x + 12 - 12x}{2x} \geq 0$$
$$\frac{x^2 - 8x + 12}{2x} \geq 0$$
$$\frac{(x-6)(x-2)}{2x} \geq 0$$
$$\frac{(x-6)(x-2)}{2x} = 0$$

when $x = 6$, $x = 2$, and is undefined when $x = 0$.

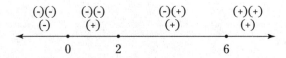

Assign signs to each of the four intervals: $x \geq 6$ or $0 < x \leq 2$

$\frac{(x-6)(x-2)}{2x}$ is undefined when $x = 0$.

21 Solve for x: $|4x - 3| < 5$. The answer is $-\frac{1}{2} < \boldsymbol{x} < \boldsymbol{2}$. First rewrite the absolute value as an inequality. Then use the rules for solving inequalities to isolate x in the middle and determine the answer:

$|4x - 3| < 5$

$-5 < 4x - 3 < 5$

$-2 < 4x < 8$, adding 3 to each section. Then, to solve for x:

$-\frac{1}{2} < x < 2$, dividing each section by 4.

Chapter 16: Working with Inequalities **203**

22 Solve for x: $|2x + 1| \geq 7$. The answer is $x \geq 3$ **or** $x \leq -4$. First, rewrite the absolute value as two inequalities. The x is isolated in the middle by adding and dividing.

$|2x + 1| \geq 7$
  $2x + 1 \geq 7$ or $2x + 1 \leq -7$
    $2x \geq 6$ or $2x \leq -8$

adding –1 to each side. Then, solve for x:

$x \geq 3$ or $x \leq -4$, dividing 2 into each side.

23 Solve for x: $|x - 3| + 6 \leq 8$. The answer is $1 \leq x \leq 5$. Before rewriting the absolute value, isolate it on one side of the inequality.

$|x - 3| + 6 \leq 8$
  $|x - 3| \leq 2$

adding –6 to each side. Now, rewrite the absolute value as an inequality that can be solved. The x gets isolated in the middle, giving you the answer.

$-2 \leq x - 3 \leq 2$
$1 \leq x \leq 5$

adding 3 to each section.

24 Solve for x: $5|7x - 4| + 1 > 6$. The answer is $x > 5/7$ **or** $x < 3/7$. Before rewriting the absolute value, it has to be alone, on the left side. First,

$5|7x - 4| + 1 > 6$
  $5|7x - 4| > 5$

adding –1 to each side. Then, $|7x - 4| > 1$, dividing 5 into each side. Now you can rewrite the absolute value as two inequality statements. Each statement is solved by performing operations that end up as x greater than or less than some value.

$7x - 4 > 1$ or $7x - 4 < -1$
  $7x > 5$ or $7x < 3$

adding 4 to each side. And, finally,

$x > 5/7$ or $x < 3/7$, dividing 7 into each side.

25 Solve for x: $6 \leq 5x + 1 < 2x + 10$. The answer is $1 \leq x < 3$. First separate $6 \leq 5x + 1 < 2x + 10$ into $6 \leq 5x + 1$ and $5x + 1 < 2x + 10$.

When $6 \leq 5x + 1$, subtract 1 from each side to get $5 \leq 5x$, and then divide each side by 5 to get $1 \leq x$.

When $5x + 1 < 2x + 10$, subtract 1 from each side and subtract $2x$ from each side to get $3x < 9$, and then divide each side by 3 to get $x < 3$. So $1 \leq x$ and $x < 3$, which gives the answer: $1 \leq x < 3$. This part of the answer includes the first part, also.

26 Solve for x: $-6 \leq 4x - 3 < 5x + 1$. The answer is $-\frac{3}{4} \leq x$ **or** $x \geq -\frac{3}{4}$. First separate $-6 \leq 4x - 3 < 5x + 1$ into $-6 \leq 4x - 3$ and $4x - 3 < 5x + 1$. Solving the first inequality, add 3 to each side to get $-3 \leq 4x$. Then divide each side by 4 to get $-\frac{3}{4} \leq x$. Solving the other inequality, you subtract 1 from each side and subtract $4x$ from each side to get $-4 < x$. This second part of the answer is an overlap of the first — all the answers are already covered in the first inequality.

For x to satisfy both inequalities, the answer is only $-\frac{3}{4} \leq x$ or $x \geq -\frac{3}{4}$.

Part IV
Applying Your Skills to Solve Story Problems

The 5th Wave — By Rich Tennant

In This Part . . .

I'm not going to beat around the bush about this part. Applications are story problems. They're one of the main reasons to become proficient at algebra. Solving an equation is great, but without some reason for it, what good is that solution? When you solve an equation and get $x = 4$, you want to be able to say, "See, I told you that the truck had four men in it!"

Chapter 17
Figuring Out Formulas

In This Chapter
▶ Using the Pythagorean theorem
▶ Introducing perimeter, area, and volume formulas
▶ Solving distance problems
▶ Computing interest and percentages

*J*ust like you use a recipe when cooking up your mother's favorite dinner for her birthday, algebra uses formulas to whip up a solution. In the kitchen you rely on the recipe that two cups of flour, ⅔ cup lard, ¾ teaspoon salt, and a ⅓ cup hot water makes a pie crust. Meanwhile, in algebra, a *formula* is an equation that expresses some relationship you can count on to help you concoct such items as a rectangle's area or the amount of interest paid on a loan.

For instance, the distance traveled is equal to the rate or speed at which you're traveling times the amount of time you travel at that speed. The formula is $d = rt$. This formula is much more compact than all those words, and, as long as you know what the letters stand for, you can use it to solve problems. You can even turn this formula around a bit and figure out how long it takes you to travel someplace if you know how far away it is and what your speed will be. You can rewrite the formula as $t = \frac{d}{r}$, which tells you to divide the distance (d) by the rate (r).

Working with formulas is easy. You can apply them to so many situations in algebra and in real life. These formulas are old, familiar friends. There's a certain comfort that comes from working with them because you know they never change (and you don't risk burning the house down like you do in the kitchen).

This chapter provides you several chances to work through some of the more common formulas and to tweak areas where you may need a little extra work. The problems are pretty straightforward, and I tell you which formula to use. In Chapter 18, you get to make decisions as to when and if to use a formula, or whether you get to come up with an equation all on your own.

Applying the Pythagorean Theorem

Pythagoras, the Greek mathematician, is credited for discovering this wonderful relationship between the lengths of the sides of a right (or 90 degree) triangle. The Pythagorean theorem is $a^2 + b^2 = c^2$. If you square the length of the two shorter sides of a right triangle and add them, then that sum equals the square of the hypotenuse's length. (The *hypotenuse* is the longest length, or the third side, in a right triangle.)

208 Part IV: Applying Your Skills to Solve Story Problems

Q. I have a helium-filled balloon attached to the end of a 500-foot string. My friend, Keith, is standing directly under the balloon, 300 feet away from me. (Oh, and yes, the ground is perfectly level, as it always is in these hypothetical situations.) These dimensions form a right triangle, with the string as the hypotenuse. My question is, how high up is the balloon?

A. **400 feet up.** Identify the parts of the right triangle in this situation, and substitute the known values into the Pythagorean theorem. Here's a picture.

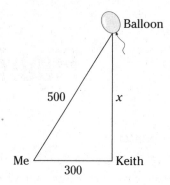

The 500-foot string forms the hypotenuse, so the equation reads $300^2 + x^2 = 500^2$. Solving for x,

$$90,000 + x^2 = 250,000$$
$$x^2 = 250,000 - 90,000 = 160,000.$$
$$x = \sqrt{160,000} = 400$$

I only use the positive solution to this equation, of course, because a negative answer doesn't make any sense.

---

1. Find the missing length in the right triangle: 10, 24, c

Solve It

2. Find the missing length in the right triangle: 15, b, 25

Solve It

---

3. A ladder from the ground to a window that's 24 feet above the ground is placed 7 feet from the base of the building, forming a right triangle. How long is the ladder, if it just reaches the window?

Solve It

4. Calista flew 400 miles due north and then turned due east and flew another 90 miles. How far is she from where she started? (*Hint:* The distance is the hypotenuse of a right triangle.)

Solve It

Getting Around Using Perimeter Formulas

The *perimeter* of something is the distance around its outside. So how do you apply perimeter? You use perimeter formulas when determining, for example, the amount of fencing you need to surround a yard, the length of a track around a field, or the amount of molding around a room along the floor.

You don't need to memorize them all, but just realize that all sorts of formulas apply for the perimeters of standard-shaped objects. These formulas are helpful for doing the needed computing, and you can alter them to solve for the desired value. Other formulas are available in geometry books, almanacs, and books of math tables.

Q. If you know that the perimeter of a particular rectangle is 20 yards and that the length is 8 yards, then what is the width?

A. **2 yards.** You can find the rectangle's perimeter by using the formula $P = 2(l + w)$ where l and w are the length and width of the rectangle. Substitute what you know into the formula and solve for the unknown. In this case, you know P and l. The formula now reads $20 = 2(8 + w)$. Divide each side of the equation by 2 to get $10 = 8 + w$. Subtract 8 from each side and you get the width, w, of 2 yards.

Q. Solve for the width of an isosceles trapezoid. (Remember, an isosceles trapezoid is a trapezoid with the two nonparallel sides equal in measure.)

A. $\dfrac{P - b_1 - b_2}{2} = w$. You can write the perimeter of an isosceles trapezoid as $P = 2w + b_1 + b_2$, where w represents the two equal sides and the two b's are the parallel bases. (For a picture of a trapezoid, look at problem 10 in this chapter.) The best approach is to solve for w and insert numbers for the perimeter and bases. To solve for w in this formula, first subtract the two bases from each side and then divide by 2:

$$P = 2w + b_1 + b_2$$
$$P - b_1 - b_2 = 2w$$
$$\dfrac{P - b_1 - b_2}{2} = w$$

5. You have 400 feet of fencing and want to fence in a rectangular yard. If the yard is to be 30 feet wide, then how long will it be?

Solve It

6. You can find the area of a square with $A = s^2$, where s represents the length of each side. Find the perimeter of a square with $P = 4s$. If the area of a square is 49 square inches, then what is its perimeter?

Solve It

7. Solve for side s in the formula for the perimeter of an isosceles triangle: $P = 2s + b$

Solve It

8. A square and an *equilateral* triangle (all three sides equal in length) have sides that are the same length. If the sum of their perimeters is 84 feet, then what is the perimeter of the square?

Solve It

Squaring Off with Area Formulas

You measure the area of a figure in square inches, square feet, square yards, and so on. Some of the more commonly found figures, such as rectangles, circles, and triangles have standard area formulas. Obscure figures even have formulas, but they aren't used very often, particularly in Algebra I and II.

Q. Find the area of a circle with a circumference of 1,256 feet.

A. **125,600 square feet**. You know that the distance around the outside (*circumference*) of a circular field is 1,256 feet. The formula for the circumference of a circle is $C = \pi d = 2\pi r$ which says that the circumference is π (about 3.14) times the diameter or two times π times the radius. To find the area of a circle, you need the formula $A = \pi r^2$. So, to find the area of this circular field, you first find the radius, by putting the 1,256 feet into the circumference formula: $1{,}256 = 2\pi r$. Replace the π with 3.14 and solve for r:

$$1256 = 2(3.14)r$$
$$\frac{1256}{6.28} = \frac{6.28r}{6.28}$$
$$200 = r$$

The radius is 200 feet. Putting that into the area formula

$$A = \pi (200)^2$$
$$A = 3.14(40{,}000) = 125{,}600$$

you get that the area is 125,600 square feet.

Q. Find the area of your bedroom.

A. **It depends**. Get out a tape measure and measure your walls. Is your bedroom a rectangle? A square? An odd shape? Use the specific formula you need. Now you know how big your kingdom is.

9. You can find the area of a rectangle with the formula $A = lw$ where l is the length and w is the width of the rectangle. What is the length of the rectangle, if you know that the area is 144 square miles and the width is 9 miles?

Solve It

10. You can find the area of a trapezoid with $A = \frac{1}{2}h(b_1 + b_2)$. Refer to the following figure and determine the length of the base b_1 if the trapezoid has an area of 30 square yards, a height of 5 yards, and the other base b_2 of 3 yards?

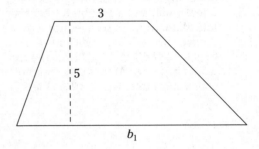

Solve It

11. The perimeter of a square is 40 feet. What is its area? (Remember: $P = 4s$ and $A = s^2$.)

Solve It

12. You can find the area of a triangle with $A = \frac{1}{2}bh$ where the base (b) and the height (h) are perpendicular to one another. If a right triangle has a base of 10 inches and a hypotenuse of 26 inches, then what is its area?

Solve It

Working with Volume Formulas

The *volume* of an object is a three-dimensional measurement. In a way, you're asking, "How many little cubes can I fit into this object?" Cubes won't fit into spheres, pyramids, or other structures with slants and curves, so you have to accept that some of these little cubes are getting shaved off or cut into pieces to fit. Having a formula is much easier than actually sitting and trying to fit all those little cubes into an often large or unwieldy object.

Q. What are the possible dimensions of a refrigerator that has a capacity of 8 cubic feet?

A. **It could be 1 foot long, 1 foot wide, and 8 feet high, or it could be 2 feet long, 2 feet wide, and 2 feet high (there are lots of answers).** A refrigerator with these dimensions isn't very efficient — or easy to get. Maybe 1 foot by 2 feet by 4 feet would be better. More likely than not, it's something more like 1½ × 2 × 2⅔ feet.

Q. Find the volume of a sugar cone that's 6 inches tall and has a radius of 1.5 inches.

A. **A little more than 14 cubic inches.** A right circular cone (that's what most traffic cones look like) has a volume that can be found if you know its radius and its height. The formula is $V = \frac{1}{3}\pi r^2 h$. As you can see, the value π is in this formula because the base is a circle. To find this cone's volume, put those values into the formula to get $V = \frac{1}{3}(3.14)(1.5^2)6 = 14.130$. You can stuff a little more than 14 cubic inches of ice cream into the cone — and then, of course, pile as much on top as you can.

13. You can find the volume of a box with $V = lwh$. Find the height if the volume is 200 cubic feet and the square base is 5 feet on each side (length and width are each 5).

Solve It

14. The volume of a sphere (ball) is $V = \frac{4}{3}\pi r^3$, where r is the radius of the sphere — the measure from the center to the outside. What is the volume of a sphere with a radius of 6 inches?

Solve It

15. You can find the volume of a right circular cylinder (soda pop can) with $V = \pi r^2 h$, where r is, of course, the radius, and h is the height of the cylinder — the distance between the two circular bases. Which has the greater volume: a cylinder with a radius of 6 cm and a height of 9 cm or a cylinder with a radius of 9 cm and a height of 4 cm?

Solve It

16. A cube is a box with the same length, width, and height. You can find its volume with the same formula as for a rectangular solid or with its own special formula $V = s^3$, where s is the length of any side. If the area of one of the sides of a cube is 36 square inches, then what is its volume? (Remember, the area of a square is $A = s^2$.)

Solve It

Distancing Yourself with the Distance Formula

The distance formula is probably the formula you're most familiar with — even though you may not think of it as using a formula all the time. The formula is $d = rt$. The d is the distance traveled, the r is the speed at which you're traveling, and the t is the amount of time spent traveling.

The only real challenge in using this formula is to be sure that the units in the different parts are the same. (For example, if the rate is in miles per hour, then you can't use the time in minutes or seconds.) If the units are different, you first have to convert them to an equivalent value before you can solve the formula.

Q. How long does it take for light from the sun to reach the earth?

A. About 8⅓ minutes. The sun is 93 million miles from the earth, and light travels at 186,000 miles per second. Using $d = rt$ and substituting the distance for d and 186,000 for r, you get 93,000,000 = 186,000 t. Dividing each side by 186,000, t comes out to be 500 seconds. Divide 500 by 60, and you get 8⅓ minutes.

Q. How fast do you go to travel 200 miles in 300 minutes?

A. Average 40 mph. Assume you're driving over the river and through the woods, you need to get to grandmother's house by the time the turkey is done, which is in 300 minutes. It's 200 miles to grandmother's house. Because your speedometer is in miles per hour, change the 300 minutes to hours by dividing by 60, which gives you 5 hours. Fill the values into the distance formula, 200 = 5r. Dividing by 5, it looks like you have to average 40 miles per hour.

17. How long will it take you to travel 600 miles if you're averaging 50 mph?

Solve It

18. What was your average rate of speed (just using the actual driving time) if you left home at noon, drove 200 miles, stopped for an hour to eat, drove another 130 miles, and arrived at your destination at 7 p.m.?

Solve It

Getting Interested in Using Percent

Percentages are a form of leveling the playing field. They're great for comparing ratios of numbers that have different bases. For instance, if you want to compare the fact that 45 men out of 80 bought the green car with the fact that 33 women out of 60 bought the green car, you can change both of these to percentages to determine who is more likely to buy the green car. (In this case, it's 56¼% men and 55% women.)

To change a ratio or fraction to a percent, divide the part by the whole and multiply by 100. For instance, in the case of the green cars, I divide 45 by 80 and get .5625. Multiplying that by 100, I get 56.25 which I write as 56¼%.

Percents also show up in interest formulas, because you earn interest on an investment or pay interest on a loan based on a percentage of the initial amount. The formula for simple interest is $I = Prt$, which is translated: Interest earned is equal to the principal invested times the interest rate (written as a decimal) times time (the number of years).

Compare the total amount of money using simple interest with the total you'd have if you invested in an account that compounded interest. *Compounding* means that the interest is added to the initial amount at certain intervals, and the interest is then figured on the new sum. The formula for compound interest is $A = P(1 + r/n)^{nt}$. The A is the total amount — the principal plus the interest earned. The P is the principal, the r is the interest rate written as a decimal, the n is the number of times each year that compounding occurs, and the t is the number of years.

Q. How much money do you have after 5 years if you invest $1,000 at 4% simple interest?

A. **$1,200**. Multiplying 1,000(.04)(5), you get that you'll earn 200 dollars in simple interest. Add that to the amount you started with for a total of $1,200.

Q. How much money will you have if you invest that same $1,000 at 4% for 5 years compounded quarterly (4 times each year)?

A. **$1,220.19**. The formula reads $A = 1,000(1 + .04/4)^{20}$. Using a calculator, it comes out to be $1,220.19. True, that's not all that much more than using simple interest, but the more money you invest, the bigger difference it makes.

19. If 60% of the class has the flu, and that 60% is 21 people, then how many are in the class?

Solve It

20. How much simple interest will you earn on $4,000 invested at 3% for 10 years? What is the total amount of money at that time?

Solve It

21. How much money will be in an account that started with $4,000 earning 3% compounded quarterly for 10 years?

Solve It

22. If you earned $500 in simple interest for an investment that was deposited at 2% interest for 5 years, how much had you invested?

Solve It

Answers to Problems on Using Formulas

The following are the answers (in bold) to the practice problems in this chapter.

1. Find the missing length in the right triangle: 10, 24, c. **c = 26**. The following figure can help.

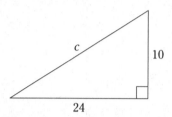

$$10^2 + 24^2 = c^2$$
$$100 + 576 = c^2$$
$$c^2 = 676 = 26^2$$
$$c = 26$$

2. Find the missing length in the right triangle: 15, b, 25. **b = 20**. Refer to the following figure.

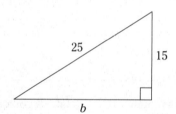

$$15^2 + b^2 = 25^2$$
$$225 + b^2 = 625$$
$$b^2 = 625 - 225 = 400 = 20^2$$
$$b = 20$$

3. A ladder from the ground to a window that's 24 feet above the ground is placed 7 feet from the base of the building. How long is the ladder, if it just reaches the window? **25 feet**. Check out the following figure.

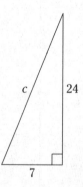

$$7^2 + 24^2 = c^2$$
$$49 + 576 = c^2$$
$$c^2 = 625 = 25^2$$
$$c = 25$$

4 Calista flew 400 miles due north and then turned due east and flew another 90 miles. How far is she from where she started? (*Hint:* The distance is the hypotenuse of a right triangle.) **410 miles.** See the following figure.

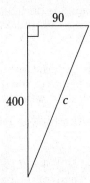

$400^2 + 90^2 = c^2$
$160,000 + 8,100 = c^2$
$c^2 = 168,100 = 410^2$ *because* $\sqrt{168,100} = 410$
$c = 410$

5 You have 400 feet of fencing and want to fence in a rectangular yard. If the yard is to be 30 feet wide, then how long will it be? **170 feet.**

Use $P = 2(l + w)$, $400 = 2(l + 30)$ because $p = 400$ and $w = 30$.

$200 = l + 30$ by dividing by 2, and $200 - 30 = l$ by subtracting 30 from each side. So $l = 170$ feet.

6 You can find the area of a square with $A = s^2$, where s represents the length of each side. Find the perimeter of a square with $P = 4s$. If the area of a square is 49 square inches, then what is its perimeter? **28 inches.** Check out this figure to help.

$A = s^2 = 49$
$s = \sqrt{49} = 7$
Perimeter: $P = 4s = 4(7) = 28$ inches

7 Solve for side s in the formula for the perimeter of an isosceles triangle: $P = 2s + b$. $\mathbf{s = \dfrac{P-b}{2}}$

Subtract b from each side to get $P - b = 2s$. Now divide by 2: $\dfrac{P-b}{2} = \dfrac{2s}{2}$ so $s = \dfrac{P-b}{2}$.

8 A square and an *equilateral* triangle (all three sides equal in length) have sides that are the same length. If the sum of their perimeters is 84 feet, then what is the perimeter of the square? **48 feet**.

The perimeter of the square is $4l$, and the perimeter of the equilateral triangle is $3l$. Adding these together, you get $4l + 3l = 7l = 84$ feet. Dividing each side of $7l = 84$ by 7, $l = 12$ feet. So the perimeter of the square is $4(12) = 48$ feet.

9 You can find the area of a rectangle with the formula $A = lw$ where l is the length and w is the width of the rectangle. What is the length of the rectangle, if you know that the area is 144 square miles and the width is 9 miles? **16 miles**.

Replace the A with 144 and the width with 9 to get $144 = l(9)$. Divide each side by 9 to get $16 = l$. The length is 16 miles.

10 You can find the area of a trapezoid with $A = \frac{1}{2}h(b_1 + b_2)$. Refer to the figure in problem 10 earlier in this chapter and determine the length of the base b_1 if the trapezoid has an area of 30 square yards, a height of 5 yards, and the other base b_2 of 3 yards? **9 yards**.

$$A = \frac{1}{2}h(b_1 + b_2)$$
$$30 = \frac{1}{2}(5)(b_1 + 3)$$
$$60 = 5(b_1 + 3)$$

multiplying each side by 2.

$12 = b_1 + 3$, dividing by 5.

$12 - 3 = b_1$ so $b_1 = 9$ yards

11 The perimeter of a square is 40 feet. What is its area? (Remember: $P = 4s$ and $A = s^2$.) **100 square feet**.

Dividing each side of $40 = 4s$ by 4, you get $10 = s$. Now find the area with $A = s^2$. Yes, 100 square feet is correct.

12 You can find the area of a triangle with $A = \frac{1}{2}bh$ where the base (b) and the height (h) are perpendicular to one another. If a right triangle has a base of 10 inches and a hypotenuse of 26 inches, then what is its area? **120 square inches**. The following figure can help.

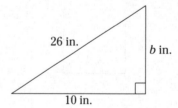

For the right triangle,

$b^2 + 10^2 = 26^2$
$b^2 + 100 = 676$
$\quad b^2 = 576 = 24^2$
$\quad b = 24$

Now find the area with $A = \frac{1}{2}bh$:

$A = \frac{1}{2}(24)(10)$
$= 12(10)$
$= 120$

13 You can find the volume of a box with $V = lwh$. Find the height if the volume is 200 cubic feet and the square base is 5 feet on each side (length and width are each 5). **8 feet**.

Replace the V with 200, and replace the l and w each with 5 to get $200 = 5(5)h = 25h$. Dividing each side of $200 = 25h$ by 25, you get $h = 8$. So the height is 8 feet.

14 The volume of a sphere (ball) is $V = \frac{4}{3}\pi r^3$, where r is the radius of the sphere — the measure from the center to the outside. What is the volume of a sphere with a radius of 6 inches? **904.32 cubic inches**.

Substituting in the 6 for r, you get $V = \frac{4}{3}(3.14)6^3 = 904.32$ cubic inches.

15 You can find the volume of a right circular cylinder (soda pop can) with $V = \pi r^2 h$, where r is, of course, the radius, and h is the height of the cylinder — the distance between the two circular bases. Which has the greater volume: a cylinder with a radius of 6 cm and a height of 9 cm or a cylinder with a radius of 9 cm and a height of 4 cm? **Neither, the volumes are the same**.

The first volume is $V = \pi(6^2)(9) = 324\pi \text{cm}^3$.

The second volume is $V = \pi(9^2)(4) = 324\pi \text{cm}^3$.

So they have the same volume, namely $324(3.14) = 1017.36$ cubic centimeters.

16 A cube is a box with the same length, width, and height. You can find its volume with the same formula as for a rectangular solid or with its own special formula $V = s^3$, where s is the length of any side. If the area of one of the sides of a cube is 36 square inches, then what is its volume? (Remember, the area of a square is $A = s^2$.) **216 cubic inches**.

So $s = \sqrt{36} = 6$ inches, and $V = s^3 = 6^3 = 216$ cubic inches

17 How long will it take you to travel 600 miles if you're averaging 50 mph? **12 hours**.

Use the distance formula ($d = rt$) and replace the d with 600 and the r with 50: $600 = 50t$. Divide each side to get $12 = t$.

18 What was your average rate of speed (just using the actual driving time) if you left home at noon, drove 200 miles, stopped for an hour to eat, drove another 130 miles, and arrived at your destination at 7 p.m.? **55 mph**.

Use the formula $d = rt$ where $d = 200 + 130 = 330$ total miles, and the time is $t = 7 - 1 = 6$ hours. Substituting into the formula, $330 = r(6)$. Dividing each side by 6 to get the average rate of speed, $55 = r$.

19 If 60% of the class has the flu, and that 60% is 21 people, then how many are in the class? **35 people**.

Let x = the number of people in the class. Then 60% of x is 21, which is written $.60x = 21$. Divide each side by .60, and you get that $x = 35$ people.

20 How much simple interest will you earn on $4,000 invested at 3% for 10 years? What is the total amount of money at that time? **$1,200 and $5,200**.

Use the formula $I = Prt$ where the principal (P) is 4,000, the rate (r) is .03, and the time (t) is 10: $I = 4,000(.03)(10) = \$1,200$. Add this amount to the original $4,000 to get a total of $5,200.

21 How much money will be in an account that started with $4,000 earning 3% compounded quarterly for 10 years? **$5,393.39**.

Use the formula for compound interest: $A = P(1 + \frac{r}{n})^{nt}$. (*Note:* Compare this total amount with the total using simple interest, in problem 20. In that problem the total is $4,000 + 1,200 = $5,200.) In the compound interest formula, $P = 4000$, $r = .03$, $n = 4$, and $t = 10$.

$$A = 4,000\left(1 + \frac{.03}{4}\right)^{4(10)}$$
$$= 4,000(1 + .0075)^{40}$$
$$= 5,393.3944$$
$$= \$5,393.39$$

So you do slightly better by letting the interest compound.

22 If you earned $500 in simple interest for an investment that was deposited at 2% interest for 5 years, how much had you invested? **$5,000**.

Use $I = Prt$ where $I = 500$, $r = .02$ and $t = 5$. Putting the numbers into the formula, $500 = P(.02)(5) = P(.1)$. Divide each side of $500 = P(.1)$ by .1, and you get that $P = 5000$, or the amount that was invested.

Chapter 18

Applying Formulas to Basic Story Problems

In This Chapter
▶ Working with perimeter, area, and volume
▶ Tackling geometry
▶ Trying distance problems

Algebra students often groan and moan when they see story problems. However, you don't need to feel any pain when you work through them. Before you grapple story problems, you need to identify what they are. I can help. You know you're facing a story problem when you see a bunch of words followed by a question.

Calm down! No need to run for the hills. The trick to doing story problems is quite simple. Change the words into a solvable equation and then answer the question based on the equation's solution. Coming up with the equation is often the greatest challenge. However, some story problems have a formula built into them to help. Look for those, first.

This chapter starts off with some of the easier story problems. These problems help by providing a built-in formula. This chapter is a good way to begin with story problems — with area, volume, and distance. They're sort of like the problems in Chapter 17, but, in this chapter, you're more on your own. You have to decide how to handle the problem and if a formula will help. If you can find a formula, just fill in the known values and you can solve for the unknown. (Check out Chapters 19 and 20 for more complex story problems.)

Deciphering Perimeter, Area, and Volume

When a problem involves perimeter, area, or volume of figures, take the formula and fill in what you know. Where do you find the formulas? One place is in this book, of course (they're sprinkled throughout the chapters). Geometry books and almanacs also have formulas. Or, you can do like my neighbors and call me. (Just not during dinnertime, please.)

For instance, the perimeter of any triangle is just the sum of the lengths of the sides. The formula reads: $P = a + b + c$. An isosceles triangle has two of the sides with equal measures, so that formula is: $P = 2s + b$.

Q. An isosceles triangle has a perimeter of 40 yards and two equal sides each 5 yards longer than the base. How long is the base?

A. The base is 10 yards long. First, you can write the triangle's perimeter as $P = 2s + b$.

The two equal sides, s, are 5 yards longer than the base, which means you can write them as $b + 5$. Putting $b + 5$ in for the s in the formula, and putting the 40 in for P, the problem now involves solving the equation $40 = 2(b + 5) + b$. I distribute the 2 to get $40 = 2b + 10 + b$. Simplifying on the right, I get $40 = 3b + 10$. Subtracting 10 from each side gives you $30 = 3b$. Dividing by 3, I get $10 = b$. So the base is 10 yards. The two equal sides are then 15 yards each. If you add the two 15-yard sides to the 10-yard base, you get (drum roll, please) $15 + 15 + 10 = 40$, the perimeter.

Q. A builder is designing a house with a square room. If she increases the sides of the room by 8 feet, the area increases by 224 square feet. What are the dimensions of those two rooms?

A. The smaller room is 10 by 10 feet; the larger room is 18 by 18 feet. You can find the area of a square with $A = s^2$, where s is the length of the sides. Start by letting the smaller room have sides measuring s feet. Its area is $A = s^2$. The larger room has sides that measure $s + 8$ feet. Its area is $A = (s + 8)^2$. The difference between these two areas is 224 feet, so subtract the smaller area from the larger, and write the equation showing the 224 as a difference: $(s + 8)^2 - s^2 = 224$. Simplify the left side of the equation:

$$s^2 + 16s + 64 - s^2 = 224$$
$$16s + 64 = 224$$

Subtract 64 from each side, and then divide by 16:

$$16s = 160$$
$$s = 10$$

1. If a rectangle is 4 inches longer than it is wide, and the area is 60 square inches, then what are the dimensions of the rectangle?

Solve It

2. A triangle has one side that's twice as long as the shortest side, and a third side that's 8 inches longer than the shortest side. Its perimeter is 60 inches. What are the lengths of the three sides?

Solve It

3. A box with a square base has a height that's three times the length of the base. If the volume is 1,536 cubic centimeters, then what is its height?

Solve It

4. If a rectangle has a length that's 3 inches greater than twice the width, and if the perimeter of the rectangle is 36 feet, then what is its length?

Solve It

5. The area of a right triangle is 330 square feet. If the height is 5 feet more than five times the base, then what is its perimeter?

Solve It

6. The volume of a cube is 216 cubic centimeters. What is the new volume if you double the length of each side?

Solve It

Using Geometry to Solve Story Problems

Geometry is a subject that has something for everyone. It has pictures, formulas, proofs, and practical applications for the homeowner. The perimeter, area, and volume formulas are considered to be a part of geometry. Geometry also deals in angle measures, parallel lines, congruent triangles, polygons and similar figures, and so forth. The different properties that have evolved from the study of geometry are helpful in solving some particular types of story problems.

Story problems using geometry are some of the more popular (if you can call any story problem popular), because they come with ready-made equations from the formulas. Also, you can draw a picture to illustrate the problem. I'm a very visual person, and I find pictures and labels on the pictures to be very helpful.

When a *transversal* (a diagonal line) cuts through two parallel lines, some interesting relations occur between the angles that are formed. It forms four *acute* (between 0 and 90 degrees) angles and four *obtuse* (between 90 and 180 degrees) angles (unless the transversal is perpendicular to the parallel lines). All the acute angles are equal, and all the obtuse angles are equal. In Figure 18-1, all the angles marked x are acute, and the angles marked y are obtuse. Also remember that the acute and obtuse angles are *supplementary* — their measures add up to 180 degrees.

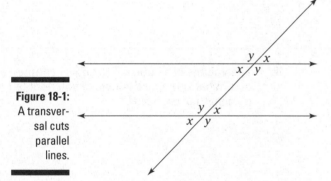

Figure 18-1: A transversal cuts parallel lines.

Q. If two parallel lines are intersected by a transversal, and the obtuse angles formed are 12 degrees greater than six times the measure of the acute angles, how big are those obtuse angles?

A. **156 degrees.** The acute and obtuse angles add up to 180 degrees, so $x + y = 180$. Replace the y with $12 + 6x$ in the equation to get $x + 12 + 6x = 180$. Combine the x terms on the left, and subtract 12 from each side to get $7x = 168$. Dividing by 7, $x = 24$. The question asks how big the obtuse angles are. The acute angles are 24, so the obtuse angles are $180 - 24 = 156$. Does that match the description, "12 degrees greater than six times the acute"? Putting the 24 in for x, $12 + 6(24) = 12 + 144 = 156$. It works.

Q. A square and *equilateral triangle* (with equal sides) have the same perimeter. The sides of the triangle are 5 inches longer than the sides of the square. What is their perimeter?

A. **60 inches.** The perimeter of the square is $4x$, where x is the length of a side. The perimeter of the triangle is $3(x + 5)$, where $x + 5$ is the length of a side (the sides are 5 inches longer than the square's). The two perimeters are equal, so $4x = 3(x + 5)$. Distribute on the right to get $4x = 3x + 15$. Subtract $3x$ from each side, and $x = 15$. Plug this into the perimeter for either the square or triangle, and you get a perimeter of 60.

7. The sum of the measures of the angles of a triangle is 180 degrees. In a certain triangle, one angle is 10 degrees greater than the smallest angle, and the biggest angle is 15 times as large as the smallest. What is the measure of that biggest angle?

Solve It

8. Two angles are *supplementary* (the sum of their measures is 180 degrees). If one of them is three times as large as the other, then what are their measures?

Solve It

9. The sum of the measures of the angles in a *quadrilateral* (a polygon with 4 sides) adds up to 360 degrees. If one of the angles is twice as big as the smallest and the other two angles are both three times as big as the smallest, then what is the measure of that smallest angle?

Solve It

10. The sum of the measures of all the angles in any polygon can be found with the formula $A = 180(n - 2)$ where n is the number of sides that the polygon has. How many sides are there on a polygon where the sum of the measures is 1,080 degrees?

Solve It

11. Two figures are *similar* when they're exactly the same shape — their corresponding angles are exactly the same measure, but they don't have to be the same size. When two figures are *similar*, their corresponding sides all have the same ratio to one another. A common way of showing that one point is the image of another (they're related some way) is to use the same letter for the second point as the first, but put a "tick" mark after the letter. In two similar triangles, the corresponding sides AB and $A'B'$ have the ratio $\frac{AB}{A'B'} = \frac{3}{4}$. In that case, what is the measure of side $C'D'$, if CD measures 45 meters?

Solve It

12. The exterior angle of a triangle lies along the same line as the interior angle it's next to. See the following figure.

The measure of an exterior angle of a triangle is always equal to the sum of the other two interior angles. If one of the non-adjacent angles in a triangle measures 30 degrees, and if the exterior angle measures 70 degrees less than twice the measure of the other nonadjacent angle, then how big is that exterior angle?

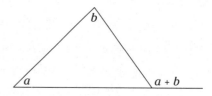

Solve It

Going the Distance with Story Problems

Distance problems use, of course, the distance formula, $d = rt$. The formula itself isn't very exciting. However, what you can do with it makes everything so very interesting.

One type of distance problem involves setting two distances equal to one another. The usual situation is that one person is traveling at one speed and another person is traveling at another speed.

Q. If Fay is traveling at an average of 40 mph and Ray is traveling at an average of 60 mph, but he left one hour later than Fay did, how far did they travel, if they traveled the same distance?

A. **120 miles**. This type of problem is where you set the distances equal. You don't know what the distance is, but you know that the rate times the time of each must equal the same thing. So, set Fay's distance, $40t$ equal to Ray's distance, $60(t - 1)$ — remember, he traveled one less hour than Fay did — and solve the equation $40t = 60(t - 1)$. Distribute the 60 on the right, giving you $40t = 60t - 60$. Subtract $60t$ from each side, resulting in $-20t = -60$. Divide by -20, and you get that $t = 3$. Fay traveled for 3 hours at 40 mph — that's 120 miles. Ray traveled for 2 hours at 60 mph, which is also 120 miles.

Q. One train leaves Kansas City traveling due east at 45 mph. A second train leaves three hours later, from Kansas City, traveling due west at 60 mph. When are they 870 miles apart?

A. **10 hours after the first train left.** This type of problem is where you add two distances together. Let the distance that the first train traveled be $45t$. This is rate times time; it equals the distance. The second train didn't travel as long; its time will be $t-3$, so represent its distance as $60(t-3)$. Add these two distances together with the sum equal to 870. $45t + 60(t-3) = 870$. Distribute the 60 on the left and simplify the terms there to get $105t - 180 = 870$. Add 180 to each side, and the equation is $105t = 1050$. Divide each side by 105 to get that $t = 10$ hours.

13. Kelly left school at 4 p.m. traveling at 25 mph. Ken left at 4:30 p.m., traveling at 30 mph, following the same route as Kelly. At what time did Ken catch up with Kelly?

Solve It

14. A Peoria Charter Coach bus left the bus terminal at 6 a.m. heading due north and traveling at an average of 45 mph. A second bus left the terminal at 7 a.m., heading due south, and traveling at an average of 55 mph. When were they 645 miles apart?

Solve It

15. Geoffrey and Grace left home at the same time. Geoffrey walked east at an average rate of 2.5 mph. Grace rode her bicycle due south at 6 mph until they were 65 miles apart. How long did it take them to be 65 miles apart?

Solve It

16. Melissa and Heather drove home for the holidays in separate cars, even though they live in the same place. Melissa's trip took two hours longer than Heather's, because Heather drove an average of 20 mph faster than Melissa's 40 mph. How far did they have to drive?

Solve It

17. Bob drove 60 mph for the first half of the time that it took for his trip, 45 mph for the next third of the time, and 66 mph for the last sixth of the time. If he drove a total of 504 miles, how long did it take him?

Solve It

18. John paddled his kayak up the Illinois River and back again to his starting point. When he paddled with the current, he traveled 40 miles in 4 hours. Against the current, he traveled the same distance in 6 hours. How fast was the current?

Solve It

Answers to Problems on Using Formulas in Story Problems

The following are the answers (in bold) to the practice problems in this chapter.

1 If a rectangle is 4 inches longer than it is wide, and the area is 60 square inches, then what are the dimensions of the rectangle? **The rectangle is 6 inches by 10 inches.** Refer to the following figure to help you solve this problem.

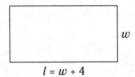

Let w = the width, which makes the length = $w + 4$. The area of a rectangle is $A = lw$. Substituting the 60 in for A and the two expressions for width and length, you get:

$60 = (w + 4)w$

$60 = w^2 + 4w$

$0 = w^2 + 4w - 60$, subtracting 60 from each side.

Factoring the quadratic, you get $0 = (w - 6)(w + 10)$

$w - 6 = 0$ when $w = 6$, or $w + 10 = 0$ when $w = -10$. But you can't have a negative width, so the width is 6 inches. And, because the length is 4 inches longer, $w + 4 = 6 + 4 = 10$ inches. The rectangle is 6 inches by 10 inches and has an area of 60 square inches.

2 A triangle has one side that's twice as long as the shortest side, and a third side that's 8 inches longer than the shortest side. Its perimeter is 60 inches. What are the lengths of the three sides? **The lengths are 13, 26, and 21 inches.** See the following figure.

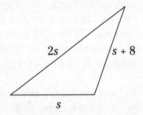

Let s = the length of the shortest side. The other two sides are $2s$ and $s + 8$. A triangle's perimeter is the sum of the lengths of the sides, so $P = 60 = s + 2s + s + 8$. Simplify to get $60 = 4s + 8$. Subtract 8 from each side: $52 = 4s$. Divide each side by 4: $s = 13$. So the three sides are 13, 2(13) = 26, and 13 + 8 = 21. As a check, add them up to see if they equal the perimeter: $13 + 26 + 21 = 60$

3 A box with a square base has a height that's three times the length of the base. If the volume is 1,536 cubic centimeters, then what is its height? **The base is 8 cm by 8 cm, and the height is 24 cm.** Refer to the following figure to help you solve this problem.

230 Part IV: Applying Your Skills to Solve Story Problems

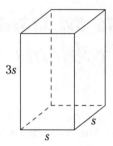

Let s = the length of an edge of the base, which makes the height $3s$ in length. The volume of a box is V = (length)(width)(height). Substituting,

$1{,}536 = (s)(s)(3s)$

$1{,}536 = 3s^3$

Dividing each side by 3, you get

$512 = s^3$

$s = \sqrt[3]{512} = 8$ cm, because $8^3 = 512$

The base is then 8 by 8, and the height is $3(8) = 24$ centimeters. To check, with the volume, 1,536 does equal 8(8)(24).

4 If a rectangle has a length that's 3 inches greater than twice the width, and if the perimeter of the rectangle is 36 feet, then what is its length? **The length is 13 inches.** Use this figure to help you solve this problem.

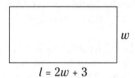

Let w = the width of the rectangle, which makes the length, $l = 3 + 2w$. A rectangle's perimeter is $P = 2(l + w)$. Substituting in for the l in this formula, and replacing P with 36, you get $36 = 2(w + 3 + 2w)$. Simplifying, you get $36 = 2(3w + 3)$. Now divide each side of the equation by 2 to get $18 = 3w + 3$. Subtract 3 from each side: $15 = 3w$. Divide each side by 3: $5 = w$. The length is $3 + 2w = 3 + 2(5) = 13$. The rectangle is 5 inches by 13 inches.

5 The area of a right triangle is 330 square feet. If the height is 5 feet more than five times the base, then what is its perimeter? **132 feet**. Refer to the following figure for guidance.

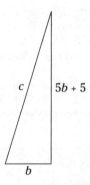

Let b = the length of the base. Then the height is $5b + 5$. Substituting them into the area formula for a triangle ($A = \frac{1}{2}bh$) you get $330 = \frac{1}{2}b(5b + 5)$. To make the equation easier, you can get rid of the fraction by multiplying each side by 2:

$660 = b(5b+5)$

$660 = 5b^2 + 5b$

Next, divide each side by 5 and move all the terms to the right by subtracting 132 from each side. Doing so gives you a quadratic equation that can be factored:

$132 = b^2 + b$

$0 = b^2 + b - 132$

$= (b+12)(b-11)$

When $b + 12 = 0$, $b = -12$. You can't have a negative length. When $b - 11 = 0$, $b = 11$. This then makes the height $5(11) + 5 = 60$. With these two sides, 11 and 60, you can solve for the length of the hypotenuse using the Pythagorean theorem ($a^2 + b^2 = c^2$): $c^2 = 11^2 + 60^2 = 121 + 3{,}600 = 3{,}721$ or $c = \sqrt{3{,}721} = 61$. The hypotenuse is 61 feet, so the perimeter is the sum of the lengths of the sides: $11 + 60 + 61 = 132$ feet

6 The volume of a cube is 216 cubic centimeters. What is the new volume if you double the length of each side? **1,728 cm³**.

First, find the lengths of the edges of the original cube. If s = the length of the edge of the cube, then use the formula for the volume of a cube ($V = s^3$):

$216 = s^3$

$s = \sqrt[3]{216} = 6$ because $6^3 = 216$

Doubling the length of the edge for the new cube results in $2(6) = 12$ centimeters. So the volume of the new cube is $V = 12^3 = 1{,}728$ cm³.

7 The sum of the measures of the angles of a triangle is 180 degrees. In a certain triangle, one angle is 10 degrees greater than the smallest angle, and the biggest angle is 15 times as large as the smallest. What is the measure of that biggest angle? **150 degrees**. Look at the figure for help.

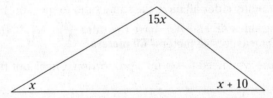

Let x = the measure of the smallest angle in degrees. Then the other two angles measure $x + 10$ and $15x$. Add them to get $x + x + 10 + 15x = 180$. Simplifying on the left, you get $17x + 10 = 180$. Subtract 10 from each side: $17x = 170$. Then, dividing by 17, $x = 10$ degrees, and the largest angle, which is 15 times as great, is 150 degrees.

To check, find the measure of the other angle, $x + 10 = 10 + 10 = 20$. Adding up the three angles, you get $10 + 20 + 150 = 180$.

8 Two angles are *supplementary* (the sum of their measures is 180 degrees). If one of them is three times as large as the other, then what are their measures? **45 and 135 degrees**. Check out the following figure.

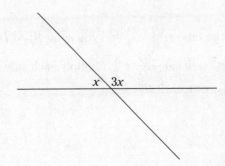

Let the measure of the smaller, acute angle be x, which makes the measure of the other angle 3x. Their sum is 180 degrees, so x + 3x = 180 give you 4x = 180. Divide each side by 4 to get x = 45 degrees. The larger is three times or 3(45)=135 degrees. 45 + 135 = 180.

9   The sum of the measures of the angles in a *quadrilateral* (a polygon with 4 sides) adds up to 360 degrees. If one of the angles is twice as big as the smallest and the other two angles are both three times as big as the smallest, then what is the measure of that smallest angle? **40 degrees**. Look at this figure.

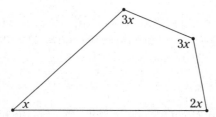

Their sum is 360, so x + 2x + 3x + 3x = 360, which simplifies to 9x = 360. Dividing by 9, x = 40 degrees. To check, find the measures of all the angles: 2x = 2(40) = 80 and 3x = 3(40) = 120. Adding them all up, you get 40 + 80 + 120 + 120 = 360 degrees.

10   The sum of the measures of all the angles in any polygon can be found with the formula A = 180(n – 2) where n is the number of sides that the polygon has. How many sides are there on a polygon where the sum of the measures is 1,080 degrees? **8 sides**.

Use the formula A = 180(n – 2), where n is that number you're trying to find. Replace the A with 1080 to get 1080 = 180(n – 2). Divide each side by 180: 6 = n – 2. Add 2 to each side, and n = 8. So it's an eight-sided polygon that has that sum for the angles.

11   Two figures are *similar* when they're exactly the same shape — their corresponding angles are exactly the same measure, but they don't have to be the same size. When two figures are *similar*, their corresponding sides all have the same ratio to one another. In two similar triangles, the corresponding sides AB and A'B' have the ratio $\frac{AB}{A'B'} = \frac{3}{4}$. In that case, what is the measure of side C'D', if CD measures 45 meters? **60 meters**.

To find the measure, you need to set up a proportion. Check out this figure to help.

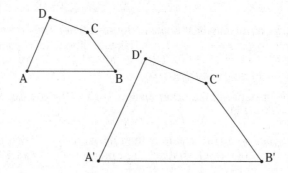

$\frac{CD}{C'D'} = \frac{AB}{A'B'}$

Because you're given the ratio $\frac{AB}{A'B'} = \frac{3}{4}$ and you know that CD = 45, you can substitute these values into the proportion to get $\frac{45}{C'D'} = \frac{3}{4}$. Multiply each side by 4: $\frac{4(45)}{C'D'} = 3$. Then, multiply each side by C'D' and divide each side by 3:

$180 = 3(C'D')$

$C'D' = \frac{180}{3} = 60$ meters

Chapter 18: Applying Formulas to Basic Story Problems

12 The measure of an exterior angle of a triangle is always equal to the sum of the other two interior angles. If one of the nonadjacent angles in a triangle measures 30 degrees, and if the exterior angle measures 70 degrees less than twice the measure of the other nonadjacent angle, then how big is that exterior angle? **130 degrees**. Check out the following figure.

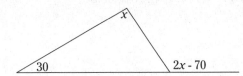

Let x = the measure of a nonadjacent angle. The other nonadjacent interior angle is 30. The exterior angle has measure $2x - 70$. Because the measure of that exterior angle is equal to the sum of the measures of the two nonadjacent interior angles, you can write the equation $2x - 70 = x + 30$. Solving for x, subtract x from each side, and add 70 to each side to get $x = 100$. The two nonadjacent interior angles add up to $30 + 100 = 130$. This amount is the same as the measure of the exterior angle. Put 100 in for x in $2x - 70$, $2(100) - 70 = 200 - 70 = 130$.

13 Kelly left school at 4 p.m. traveling at 25 mph. Ken left at 4:30 p.m., traveling at 30 mph, following the same route as Kelly. At what time did Ken catch up with Kelly? **7 p.m.**

Let t = time in hours after 4 p.m. Kelly's distance is $25t$, and Ken's distance is $30(t - ½)$ using $t - ½$ because he left a half hour after 4 p.m.

Ken will overtake Kelly when their distances are equal. So the equation to use is $25t = 30(t - ½)$. Distributing on the right, $25t = 30t - 15$. Subtract $30t$ from each side to get $-5t = -15$. Divide each side by -5 and $t = 3$. If t is the time in hours after 4 p.m., then $4 + 3 = 7$, or Ken caught up with Kelly at 7 p.m.

14 A Peoria Charter Coach bus left the bus terminal at 6 a.m. heading due north and traveling at an average of 45 mph. A second bus left the terminal at 7 a.m., heading due south, and traveling at an average of 55 mph. When were they 645 miles apart? **1 p.m.**

Let t = time in hours after 6 a.m. The distance the first bus traveled is $45t$, and the distance the second bus traveled is $55(t - 1)$, which represents 1 hour less of travel time. The sum of their distances traveled gives you their distance apart. So $45t + 55(t - 1) = 645$. Distribute the 55: $45t + 55t - 55 = 645$. Combine the like terms on the left, and add 55 to each side: $100t = 700$. Divide by 100, $t = 7$. They were 645 miles apart 7 hours after the first bus left. Add 7 hours to 6 a.m., and you get 1 p.m.

15 Geoffrey and Grace left home at the same time. Geoffrey walked east at an average rate of 2.5 mph. Grace rode her bicycle due south at 6 mph until they were 65 miles apart. How long did it take them to be 65 miles apart? **10 hours**.

Look at the following figure, showing the distances and directions that Geoffrey and Grace traveled.

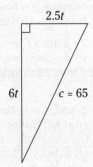

Geoffrey walks $d = rt = 2.5t$ miles, and Grace rides $d = rt = 6t$ miles. Use the Pythagorean theorem:

$$(6t)^2 + (2.5t)^2 = 65^2$$
$$36t^2 + 6.25t^2 = 65^2$$

Multiply each side by 4 to get rid of the decimal:

$$144t^2 + 25t^2 = 4 \times 65^2$$
$$169t^2 = 4 \times 65^2$$
$$t^2 = \frac{4 \times 65^2}{169}$$
$$t = \sqrt{\frac{4 \times 65^2}{169}} = \frac{2 \times 65}{13} = \frac{130}{13} = 10 \text{ hours}$$

In 10 hours, Geoffrey will walk 2.5(10) = 25 miles, and Grace will ride 6(10) = 60 miles. Plug it into the Pythagorean theorem: $25^2 + 60^2 = 625 + 3{,}600 = 4{,}225 = 65^2$.

16 Melissa and Heather drove home for the holidays in separate cars, even though they live in the same place. Melissa's trip took two hours longer than Heather's, because Heather drove an average of 20 mph faster than Melissa's 40 mph. How far did they have to drive? **240 miles**.

Because Heather's time is shorter, let t = Heather's time in hours. Heather's distance is $60t$, using 40 + 20 for her speed. Melissa's distance is $40(t + 2)$, because she took 2 hours longer. Their distances are equal, so $60t = 40(t + 2)$. Distribute the 40: $60t = 40t + 80$. Subtract $40t$ from each side: $20t = 80$. Divide by 20 for $t = 4$. Heather's distance is $60(4) = 240$ miles. Melissa's distance is $40(4 + 2) = 40(6) = 240$ miles. It's the same, of course.

17 Bob drove 60 mph for the first half of the time that it took for his trip, 45 mph for the next third of the time, and 66 mph for the last sixth of the time. If he drove a total of 504 miles, how long did it take him? **9 hours**.

Let t = total time to make the 504-mile trip. The distance traveled at 60 mph is $60\left(\frac{t}{2}\right)$. At 45 mph it's $40\left(\frac{t}{3}\right)$, and at 66 mph it's $66\left(\frac{t}{6}\right)$. The sum of these three distances will be 504 miles:

$$60\left(\frac{t}{2}\right) + 45\left(\frac{t}{3}\right) + 66\left(\frac{t}{6}\right) = 504$$
$$30t + 15t + 11t = 504$$
$$56t = 504$$
$$t = \frac{504}{56} = 9 \text{ hours}$$

Checking the work, at 60 mph, the time would be 4.5 hours and the distance 270 miles. At 45 mph, the time would be 3 hours and the distance 135 miles. At 66 mph, the time is 1.5 hours and the distance 99 miles. Add up the distances: 270 + 135 + 99 = 504.

18 John paddled his kayak up the Illinois River and back again to his starting point. When he paddled with the current, he traveled 40 miles in 4 hours. Against the current, he traveled the same distance in 6 hours. How fast was the current? **5/3 mph**.

Let s = the current's speed, and let r = John's speed in still water. Using these variables, John's speed going downstream is $r + s$. Use the distance formula ($d = rt$): $40 = (r + s)(4)$. Divide each side by 4 to get $10 = r + s$. Subtract s from each side to get $10 - s = r$. Use this equation in the next part of the solution.

John's speed going upstream is $r - s$, and, using the distance formula again, $40 = (r - s)(6)$. Now replace the r with $10 - s$, which you already simplified. You get $40 = (10 - s - s)(6)$, which simplifies to $40 = (10 - 2s)(6)$. Distribute the 6: $40 = 60 - 12s$. Subtract 60 from each side, and the equation becomes $-20 = -12s$. Divide each side by -12, and $s = -20/-12 = 5/3$ mph. This is the current's speed. To find John's speed, $r = 10 - s = 10 - 5/3 = 25/3$.

Going downstream, John's speed is $r + s = 25/3 + 5/3 = 30/3 = 10$ mph because d = (downstream speed)(downstream time) = (10)(4) = 40, which is the distance.

Chapter 19
Comparing Things in Story Problems

In This Chapter
▶ Dealing with age problems
▶ Counting up consecutive integers
▶ Working on sharing the workload

I'm sorry, but as much as you may dread story problems, you can't escape them in algebra. They're kind of like that recurring nightmare you had as a child. It just keeps rearing its ugly head, again and again. However, you don't have to stay hidden under the covers. Chapter 18 introduces you to the often-despised story problem, and now this chapter helps you look that story problem in the eye without any fear. Specifically in this chapter I focus on age problems, consecutive integers, and work problems.

Problems dealing with age and consecutive integers have something in common. You use one or more base values or ages and keep adding the same number to each. As you work through these problems, you can find some patterns in numbers, and the theme is to use the same format — pick a variable for a number and add something on to it.

As for work problems, they're a completely different bird. In work problems you need to divvy up the work and add it all together in order to get the whole job done. The portions aren't equal.

This chapter offers plenty of opportunities for you to tackle these story problems and overcome any trepidation.

Answering Age Problems

Age problems in algebra don't have anything to do with wrinkles or thinning hair. Algebra deals with age problems very systematically and with an eye to the future. A story problem involving ages usually includes something like "in four years" or "ten years ago." The trick is to incorporate the information in the story problem into an equation.

When establishing your equation, make sure you understand and keep track of how you name your variables. The letter x doesn't stand for Joe. The letter x can stand for Joe's *age*. If you keep in mind that the variables stand for numbers, it all makes more sense.

Q. Joe's father is twice as old as Joe is. (He hasn't always been and won't be again — think about it: When Joe was born, was his father twice as old as he was?) Twelve years ago, Joe's father was three years older than three times Joe's age. How old is Joe's father now?

A. Joe's father is **42 years old**.

1. **Assign a variable to Joe's age.**

   Let Joe's age be x. Joe's father is twice as old, so Joe's father's age is $2x$.

2. **Continue to read through the problem.**

   Twelve years ago... Both Joe's age and his father's age have to be backed up by 12 years. Their respective ages, 12 years ago, are $x - 12$ and $2x - 12$.

3. **Take the rest of the sentence where "Joe's father was three years..." and change it into an equation, putting an equal sign where the verb is.**

   "Twelve years ago, Joe's father" becomes $2x - 12$

   "was" becomes =

   "three years older than" becomes 3 +

   "three times Joe's age (twelve years ago)" becomes $3(x - 12)$

4. **Put this information all together.**

   $2x - 12 = 3 + 3(x - 12)$

5. **Solve for x.**

   $2x - 12 = 3x - 33$, and $x = 21$

   That's Joe's age, so his father is twice that or 42.

---

1. Jack is three times as old as Chloe. Ten years ago, Jack was five times as old as Chloe. How old are Jack and Chloe now?

Solve It

2. Linda is ten years older than Luke. In ten years, Linda's age will be thirty years less than twice Luke's age. How old is Linda now?

Solve It

3. Avery is six years older than Patrick. In four years, the sum of their ages will be 26. How old is Patrick now?

Solve It

4. Jon is three years older than Jim, and Jim is two years older than Jane. Ten years ago, the sum of their ages was 40. How old is Jim now?

Solve It

Tackling Consecutive Integer Problems

When items are *consecutive,* they follow along after one another. An *integer* is a positive or negative whole number, or 0. So, put these two things together to get *consecutive integers.* Consecutive integers have patterns — they're evenly spaced. The following three lists are examples of consecutive integers:

Consecutive integers: 5, 6, 7, 8, 9, . . .

Consecutive odd integers: 11, 13, 15, 17, 19, . . .

Consecutive multiples of 8: 48, 56, 64, 72, 80, . . .

After you get one of the integers in a list, you can pretty much get all the rest. When doing consecutive integer problems, let one of the integers (usually the first in your list) be represented by x, and then add on 1, 2, or whatever the spacing is to the next number, and then add that on to the new number, and so on, depending on how many integers you need.

Q. The sum of six consecutive integers is 255. What are they?

A. The integers are **40, 41, 42, 43, 44 and 45**. The first integer in my list is x. The next is $x + 1$, the one after that is $x + 2$, and so on. Replacing the x with 40 in each case gives you all these answers. The equation for this situation reads $x + (x + 1) + (x + 2) + (x + 3) + (x + 4) + (x + 5) = 255$. The parentheses aren't necessary. I just include them so you can see the separate terms. Adding up all the x's and numbers, the equation becomes $6x + 15 = 255$. Subtracting 15 from each side and dividing each side by 6, I get $x = 40$. Fill the 40 into your original equations to get the six consecutive integers. (If $x = 40$, then $x + 1 = 41$, $x + 2 = 42$, and so on.)

Q. The sum of four consecutive odd integers is 8. What are they?

A. The integers are **–1, 1, 3, and 5**. The first number is –1. Replace the x with –1 in each case to get the rest of the answers. The equation for this problem is $x + (x + 2) + (x + 4) + (x + 6) = 8$. It becomes $4x + 12 = 8$. Subtracting 12 and then dividing by 4, $x = -1$. You may have questioned using the +2, +4, +6 when dealing with odd integers. The problem designates x as an odd integer, and the other integers are all two steps away from one another. It works!

5. The sum of three consecutive integers is 57. What are they?

Solve It

6. The sum of four consecutive even integers is 52. What is the largest of the four?

Solve It

7. The sum of three consecutive odd integers is 75. What is the middle number?

Solve It

8. The sum of five consecutive multiples of 4 is 20. What are they?

Solve It

9. The sum of the smallest and largest of three consecutive integers is 126. What is the middle number of those consecutive integers?

Solve It

10. The product of two consecutive integers is 89 more than their sum. What are they?

Solve It

Working Together on Work Problems

Work problems in algebra involve doing jobs alone and together. Together is usually better, unless the person you're working with distracts you. I take the positive route and assume that two heads are better than one.

The general format for these problems is to let x represent how long it takes to do the job working together. Follow these steps and you won't even break a sweat when working work problems.

1. **Write the amount that a person can do in one time period as a fraction.**
2. **Multiply that amount by the x.**

    (You've multiplied each fraction that each person can do by the time it takes to do the whole job.)

3. **Add the portions of the job that are completed in one time period together and set the sum equal to 1.**

    (Setting the amount to 1 is 100% of the job.)

Q. Meg can clean out the garage in five hours. Mike can clean out the same garage in three hours. How long will it take if they work together?

A. **Slightly less than two hours.** Let x represent the amount of time it takes to do the cleaning when they're working together. Meg can do $\frac{1}{5}$ of the job in one hour, and Mike can do $\frac{1}{3}$ of the job in one hour. The equation to use is $\frac{1}{5}x + \frac{1}{3}x = 1$. Multiply both sides by the common denominator, and add the two fractions together to get $15\left(\frac{1}{5}x + \frac{1}{3}x\right) = 15(1)$. Now just multiply by the reciprocal of 8, and you get

$$3x + 5x = 15$$
$$8x = 15$$
$$x = \frac{15}{8}$$

Q. Carlos can wash the bus in two hours, and Carol can wash the bus in 20 hours. How long will it take if they work together?

A. **Slightly less than two hours; it hardly seems worth getting Carol involved!** Let x represent the amount of time it takes to wash the bus when they're working together. Carlos can do $\frac{1}{2}$ of the job in one hour, and Carol can do $\frac{1}{20}$ of the job in one hour. The equation to use is $\frac{x}{2} + \frac{x}{20} = 1$. Multiply both sides by 20, and you get $10x + x = 20$. Adding the two terms on the left together, you get $11x = 20$. Dividing by 11, $x = \frac{20}{11}$.

11. Alissa can do the job in three days, and Alex can do the same job in four days. How long will it take if they work together?

Solve It

12. George can paint the garage in five days, Geanie can paint it in eight days, and Greg can do the job in ten days. How long will it take if they all work together?

Solve It

13. Working together, Sam and Helene wrote a company organizational plan in $1\frac{1}{3}$ days. Working alone, it would have taken Sam four days to write that plan. How long would it have taken Helene, if she had written it alone?

Solve It

14. Rancher Biff needs his new fence put up in four days — before the herd arrives. Working alone, it'll take him six days to put up all the fencing. He can hire someone to help. How fast does the hired hand have to work in order for the team to complete the job before the herd arrives?

Solve It

Answers to Problems on Comparing Things in Story Problems

The following are the answers (in bold) to the practice problems in this chapter.

1 Jack is three times as old as Chloe. Ten years ago, Jack was five times as old as Chloe. How old are Jack and Chloe now? **Chloe is 20, and Jack is 60**.

Let x = Chloe's present age and $3x$ = Jack's present age. Ten years ago, Chloe's age was $x - 10$, and Jack's was $3x - 10$. Also, ten years ago, Jack's age was 5 times Chloe's age. You can write this equation as $3x - 10 = 5(x - 10)$. Distribute the 5 to get $3x - 10 = 5x - 50$. Subtract $3x$ from each side, and add 50 to each side, and the equation becomes $40 = 2x$. Divide by 2 to get $x = 20$.

2 Linda is ten years older than Luke. In ten years, Linda's age will be thirty years less than twice Luke's age. How old is Linda now? **Luke is 30, and Linda is 40. In ten years, Luke will be 40, and Linda will be 50**.

Let x = Luke's age now. That makes Linda's present age = $x + 10$. In 10 years, Luke will be $x + 10$, and Linda will be $(x + 10) + 10 = x + 20$. But at these new ages, Linda's age will be 30 years less than twice Luke's age. This is written: $x + 20 = 2(x + 10) - 30$. Distributing the 2, $x + 20 = 2x + 20 - 30$. Simplifying, $x + 20 = 2x - 10$. Subtract x from each side, and add 10 to each side to get $x = 30$. Luke is 30, and Linda is 40. In ten years, Luke will be 40, and Linda will be 50. Twice Luke's age then, minus 30 is $80 - 30 = 50$. It checks.

3 Avery is six years older than Patrick. In four years, the sum of their ages will be 26. How old is Patrick? **Patrick is 6**.

Let x = Patrick's age now. Then Avery's age = $x + 6$. In 4 years, Patrick will be $x + 4$ years old, and Avery will be $x + 6 + 4 = x + 10$ years old. To write that the sum of their ages in four years: $(x + 4) + (x + 10) = 26$. Simplify on the left to get $2x + 14 = 26$. Subtract 14 from each side: $2x = 12$. Divide by 2, and you get $x = 6$. Patrick is 6, and Avery is 12. In four years, Patrick will be 10, and Avery will be 16. The sum of 10 and 16 is 26.

4 Jon is three years older than Jim, and Jim is two years older than Jane. Ten years ago, the sum of their ages was 40. How old is Jim now? **Jim is 23**.

Let x = Jane's age now. Jim is two years older, so Jim's age = $x + 2$. Jon is three years older than Jim, so Jon's age = $(x + 2) + 3 = x + 5$. Ten years ago their ages were: Jane's $x - 10$, Jim's $x + 2 - 10 = x - 8$, and Jon's $x + 5 - 10 = x - 5$. Add the ages ten years ago together to get 40: $x - 10 + (x - 8) + (x - 5) = 40$. Simplifying on the left, $3x - 23 = 40$. Add 23 to each side to get $3x = 63$. Divide by 3, and $x = 21$. Jane's age is 21, so Jim's age is $21 + 2 = 23$.

5 The sum of three consecutive integers is 57. What are they? **The integers are 18, 19, and 20**.

Let x = the smallest of the three consecutive integers. Then the other two are $x + 1$ and $x + 2$. Adding the integers together to get 57, $x + (x + 1) + (x + 2) = 57$, which simplifies to $3x + 3 = 57$. Subtract 3 from each side to get $3x = 54$. Dividing by 3, $x = 18$. $18 + 19 + 20 = 57$.

6 The sum of four consecutive even integers is 52. What is the largest of the four? **16. The four integers are 16, 14, 12, and 10**.

Let that largest integer = x. The other integers will be 2, 4, and 6 smaller, so they can be written with $x - 2$, $x - 4$, and $x - 6$. Adding them, $x + (x - 2) + (x - 4) + (x - 6) = 52$. Simplifying gives you $4x - 12 = 52$. Add 12 to each side to get $4x = 64$. Divide each side by 4, and $x = 16$. $16 + 14 + 12 + 10 = 52$.

7 The sum of three consecutive odd integers is 75. What is the middle number? **25.**

You can add and subtract 2 from that middle integer. Let that integer = x. Then the other two are $x + 2$ and $x - 2$. Add them together: $(x - 2) + x + (x + 2) = 75$. Simplifying, $3x = 75$. Divide by 3 to get $x = 25$. That's the middle number. The other two are 23 and 27: 23 + 25 + 27 = 75.

8 The sum of five consecutive multiples of 4 is 20. What are they? **The numbers are –4, 0, 4, 8, and 12.**

Let x = the first of the consecutive multiples of 4. Then the other four are $x + 4$, $x + 8$, $x + 12$, and $x + 16$. Add them together: $x + (x + 4) + (x + 8) + (x + 12) + (x + 16) = 20$. Simplifying on the left, $5x + 40 = 20$. Subtract 40 from each side to get $5x = -20$. Dividing by 5, $x = -4$.

9 The sum of the smallest and largest of three consecutive integers is 126. What is the middle number of those consecutive integers? **63.**

Let x = the smallest of the consecutive integers. Then the other two are $x + 1$ and $x + 2$. Because the sum of the smallest and largest is 126, you can write it as $x + (x + 2) = 126$. Simplifying the equation, $2x + 2 = 126$. Subtract 2 from each side to get $2x = 124$. Dividing by 2, $x = 62$, which is the smallest integer. The middle one is one bigger, so it's 63.

10 The product of two consecutive integers is 89 more than their sum. What are they? **The numbers are 10 and 11 or –9 and –8.**

Let the smaller of the integers = x. The other one is then $x + 1$. Their product is written $x(x + 1)$ and their sum is $x + (x + 1)$. Now, to write that their product is 89 more than their sum, the equation is $x(x + 1) = 89 + x + (x + 1)$. Distributing the x on the left and simplifying on the right, $x^2 + x = 2x + 90$. Subtract $2x$ and 90 from each side to set the quadratic equation equal to 0: $x^2 - x - 90 = 0$. The trinomial on the left side of the equation factors to give you $(x - 10)(x + 9) = 0$. $x = 10$ or $x = -9$. If $x = 10$, then $x + 1 = 11$. The product of 10 and 11 is 110. That's 89 bigger than their sum, 21. What about if $x = -9$? The next bigger number is then –8. Their product is 72. The difference between their product of 72 and sum of –17 is 72 –(–17) = 89. So this problem has two possible solutions.

11 Alissa can do the job in three days, and Alex can do the same job in four days. How long will it take if they work together? **It will take a little less than 2 days working together ($^{12}/_7$ days).**

Let x = the number of days to do the job together. Alissa can do $1/3$ of the job in one day, and $\frac{1}{3}(x)$ of the job in x days. In x days, as they work together, they're to do 100% of the job.

$\frac{1}{3}(x) + \frac{1}{4}(x) = 1$, which is 100%.

Multiply by 12: $4x + 3x = 12$, or $7x = 12$. Divide by 7: $x = 12/7 = 1\,5/7$ days to do the job. Alissa's share is $1/3(12/7) = 4/7$, and Alex's share is $1/4(12/7) = 3/7$. Together $4/7 + 3/7 = 1$.

12 George can paint the garage in five days, Geanie can paint it in eight days, and Greg can do the job in ten days. How long will it take if they all work together? **$2\,6/17$ days.**

Let x = the number of days to complete the job together. In x days, George will paint $\frac{1}{5}(x)$ of the garage, Geanie $\frac{1}{8}(x)$ of the garage, and Greg $\frac{1}{10}(x)$ of the garage. The equation for completing the job is $\frac{1}{5}(x) + \frac{1}{10}(x) + \frac{1}{8}(x) = 1$. Multiplying through by 40, which is the least common denominator of the fractions, $8x + 4x + 5x = 40$. Simplifying, you get $17x = 40$. Dividing by 17, $x = 40/17 = 2\,6/17$ days to paint the garage. Checking the answer, George's share is $1/5(40/17) = 8/17$, Geanie's share is $1/8(40/17) = 5/17$, and Greg's share is $1/10(40/17) = 4/17$. Together, $8/17 + 5/17 + 4/17 = 17/17$ or 100%.

13 Working together, Sam and Helene wrote a company organizational plan in $1\,1/3$ days. Working alone, it would have taken Sam four days to write that plan. How long would it have taken Helene, if she had written it alone? **2 days.**

Let x = the number of days for Helene to write the plan alone. So Helene writes $\frac{1}{x}$ of the plan each day. Sam writes $\frac{1}{4}$ of the plan per day. In $1\frac{1}{3} = \frac{4}{3}$ days, they complete the job together. The equation is $(\frac{1}{x})(\frac{4}{3}) + (\frac{1}{4})(\frac{4}{3}) = 1$. Multiplying each side by $12x$, the least common denominator, $\frac{1}{x}(12x) + \frac{1}{4}(12x) = (1)(12x)$, $16 + 4x = 12x$. Subtract $4x$ from each side to get $16 = 8x$. Dividing by 8, $x = 2$ days. Helene will complete the job in 2 days. To check this, in $\frac{4}{3}$ days, Sam will do $\frac{1}{4}(\frac{4}{3}) = \frac{1}{3}$ of the work and Helene will do $\frac{1}{2}(\frac{4}{3}) = \frac{2}{3}$. Together, $\frac{1}{3} + \frac{2}{3} = 1$ or 100%.

14. Rancher Biff needs his new fence put up in four days — before the herd arrives. Working alone, it'll take him six days to put up all the fencing. He can hire someone to help. How fast does this person have to work in order for him or her to complete the job before the herd arrives? **The hired hand must be able to do the job alone in 12 days.**

Let x = the number of days the hired hand needs to complete the job alone. The hired hand does $\frac{1}{x}$ of the fencing each day, and Biff puts up $\frac{1}{6}$ of the fence each day. In four days, they can complete the project together, so $(\frac{1}{x})(4) + (\frac{1}{6})(4) = 1$. Multiply by the common denominator $6x$: $(\frac{1}{x})(6x) + (\frac{1}{6})(6x) = (1)(6x)$. Simplifying, $24 + 4x = 6x$. Subtracting $4x$ from each side, $24 = 2x$. Dividing by 2, $x = 12$. The hired hand must be able to do the job alone in 12 days. To check this, in four days, Biff does $(\frac{1}{6})(4) = \frac{2}{3}$ of the fencing, and the hired hand $(\frac{1}{12})(4) = \frac{1}{3}$ of the job.

Chapter 20

Weighing In on Quality and Quantity Story Problems

In This Chapter
▶ Jumbling it up with mixtures
▶ Understanding the strength of a solution
▶ Counting on money in mixtures and interest problems

The story problems in this chapter have a common theme to them — they deal with quality and quantity and adding up to a total amount. (Chapters 18 and 19 have other types of story problems.) You encounter quantity and quality problems almost on a daily basis. For instance, if you have four dimes, you know that you have forty cents. How do you know? You multiply the *quantity*, four dimes, times the *quality*, ten cents each, to get the total amount of money.

In this chapter, take time to practice with these story problems. Just multiply the amount of something, the *quantity*, times the strength or worth of it, *quality*, in order to solve them.

Achieving the Right Blend with Mixtures Problems

Mixtures include what goes in granola, blends of coffee, or the colors of sugarcoated candies in a candy dish. In these types of problems, you often need to find some sort of relationship about the mixture. Put the information into an equation and solve. Just be sure that the variable represents some number — an amount or value. Bring along an appetite. Most of these problems deal with food.

Q. A health store is mixing up some granola that has many ingredients, but three of the basics are oatmeal, wheat germ, and raisins. Oatmeal costs $1 per pound, wheat germ costs $3 per pound, and raisins cost $2 per pound. The store wants to create a mixture of those three ingredients that will cost $1.50 per pound. (This amount is the base; the rest of the ingredients and additional cost will be added later.) The granola is to have nine times as much oatmeal as wheat germ. How much of each ingredient is needed?

A. **The granola will need $\frac{1}{16}$ pound of wheat germ, $\frac{9}{16}$ pound of oatmeal and $\frac{6}{16}$ or $\frac{3}{8}$ pound of raisins per pound of mixed granola.** To start this problem, let x represent the amount of wheat germ in pounds. Because you need nine times as much oatmeal as wheat germ, you'll have $9x$ pounds of oatmeal. How much in raisins? The raisins can have whatever's left of the pound after the wheat germ and oatmeal are taken out — that's $1 - (x + 9x)$ or $1 - 10x$ pounds. Now, multiply each of these amounts by their respective price: $3(x) + 1(9x) + 2(1 - 10x)$. Set this equal to the $1.50 price multiplied by its amount, as follows: 1 pound: $3(x) + 1(9x) + 2(1 - 10x) = 1.50(1)$. Simplify and solve for x:

$$3x + 9x + 2 - 20x = 1.50$$
$$-8x = -0.5$$
$$x = \frac{-0.5}{-8} = \frac{\frac{1}{2}}{8} = \frac{1}{16}$$

1. Kathy's Kandies features a mixture of chocolate creams and chocolate-covered caramels that sells for $6 per pound. If creams sell for $4.50 per pound, and caramels sell for $7 per pound, how much of each type of candy should be in a one-pound mix?

Solve It

2. Solardollars Coffee is trying new blends to attract more customers. The premium Colombian costs $10 per pound, and the regular blend costs $4 per pound. How much of each should the company use to make 100 pounds of a coffee blend that costs $5.50 per pound?

Solve It

3. Peanuts cost $2 per pound, almonds cost $3.50 per pound, and cashews cost $6 per pound. How much of each should you use to create a mixture that costs $3.40 per pound, if you have to use twice as many peanuts as cashews?

Solve It

4. A mixture of jellybeans is to contain twice as many red as yellow, three times as many green as yellow, and twice as many pink as red. How many of each color jellybean should be in a bag of 100 jellybeans?

Solve It

5. A *Very Berry Smoothie* calls for raspberries, strawberries, and yogurt. Raspberries cost $3 per cup, strawberries cost $1 per cup, and yogurt costs $0.50 per cup. The recipe calls for twice as much strawberries as raspberries. How many cups of strawberries are needed to make a gallon of this smoothie that costs $10.10? (*Hint:* 1 gallon = 16 cups)

Solve It

6. A supreme pizza contains five times as many ounces of cheese as mushrooms, twice as many ounces of peppers as mushrooms, twice as many ounces of onions as peppers, and four more ounces of sausage than mushrooms. If the topping on this pizza is to weigh 30 ounces, then how many ounces of each ingredient is used?

Solve It

Finding the Correct Solution One Hundred Percent of the Time

Solutions problems are sort of like mixtures problems (see "Achieving the Right Blend with Mixtures Problems" earlier in this chapter). The main difference is that solutions usually deal in percents — 30%, 27½%, 0%, or even 100%. These last two numbers indicate, respectively, that none of that ingredient is in the solution (0%) or that it's *pure* for that ingredient (100%). You've dealt with these solutions if you've had to add antifreeze or water to your radiator. Or how about mixing up that new punch with orange juice and ginger ale?

The general format for these problems is

(percent A × amount A) + (percent B × amount B) = (percent C × amount C)

Q. How many gallons of 60% apple juice mix need to be added to 20 gallons of mix that's currently 25% apple juice to bring it up to a mix that's 32% apple juice?

A. **Five gallons.** To solve this problem, let x represent the unknown amount of 60% apple juice. Using the format of all the percents times the respective amounts, you get $(60\% \times x) + (25\% \times 20 \text{ gallons}) = (32\% \times (x + 20 \text{ gallons}))$. Change the percents to decimals and solve the equation:

$$.60x + .25(20) = .32(x + 20)$$
$$.60x + 5 = .32x + 6.4$$
$$.28x = 1.4$$
$$x = 5$$

If you don't care for decimals, you could multiply each side by 100 to change everything to whole numbers. If you're adding pure alcohol or pure antifreeze or something like this, use 1 (which is 100%) in the equation. If there's no alcohol, chocolate syrup, salt, or whatever in the solution, use 0 (which is 0%) in the equation.

7. How many quarts of 25% solution do you need to add to 4 quarts of 40% solution to create a 31% solution?

Solve It

8. How many gallons of 5% solution have to be added to 2 gallons of 90% solution to create 15% solution?

Solve It

9. How many quarts of pure antifreeze need to be added to 8 quarts of 30% antifreeze to bring it up to 50%?

Solve It

10. How many cups of chocolate syrup need to be added to 1 quart of milk to get a mixture that's 25% syrup?

Solve It

11. What concentration should the 4 quarts of salt water have so that when it's added to 5 quarts of 40% solution salt water the concentration goes down to $33\frac{1}{3}$%?

Solve It

12. What concentration and amount solution have to be added to 7 gallons of 60% alcohol to produce 16 gallons of $37\frac{1}{2}$% alcohol solution?

Solve It

Watching Your Pennies with Money Problems

Story problems involving coins, money, or interest earned all involve a process like that in solutions problems — you multiply a quantity times a quality. In these cases, the qualities are the values of the coins or bills, or they're the interest rate at which money is growing.

Q. Gabriella is counting the bills in her cash drawer before going to work for the day. She has the same number of $10 bills as $20 bills. She has two more $5 bills than $10 bills, and ten times as many $1 bills as $5 bills. She has a total of $300 in bills. How many of each does she have?

A. **Gabriella has 6 $10 bills, 6 $20 bills, 8 $5 bills, and 80 $1 bills for a total of $300.** You can compare everything, directly or indirectly, to the $10 bills. Let x represent the number of $10 bills. The number of $20 bills is the same. The number of $5 bills is two more than the number of $10 bills, so let that be represented by $x + 2$. Multiply $x + 2$ by 10 for the number of $1 bills, $10(x + 2)$. Now take each *number* of bills and multiply by the quality or value of that bill. Add them to get $300:

$$10(x) + 20(x) + 5(x+2) + 1(10(x+2)) = 300$$
$$10x + 20x + 5x + 10 + 10x + 20 = 300$$
$$45x + 30 = 300$$
$$45x = 270$$
$$x = 6$$

Q. Jon won the state lottery and has one million dollars to invest. He invests some of it in a highly speculative venture that earns 18% interest. The rest is invested more wisely, at 5% interest. If he earns $63,000 in simple interest in one year, how much did he invest at 18%?

A. **He invested $100,000 at 18% interest.** Let x represent the amount of money invested at 18%. Then the remainder, $1,000,000 - x$, is invested at 5%. The equation to use is $.18x + .05(1,000,000 - x) = 63,000$. Distributing the .05 on the left and combining terms, you get $.13x + 50,000 = 63,000$. Subtract 50,000 from each side, and $.13x = 13,000$. Dividing each side by .13, $x = 100,000$. It's kind of mind boggling.

13. Carlos has twice as many quarters as nickels and has a total of $8.25. How many quarters does he have?

Solve It

14. Gregor has twice as many $10 bills as $20 bills, five times as many $1 bills as $10 bills, and half as many $5 bills as $1 bills. He has a total of $750. How many of each bill does he have?

Solve It

15. Stella has 100 coins in nickels, dimes, and quarters. She has 18 more nickels than dimes and a total of $7.40. How many of each coin does she have?

Solve It

16. Betty invested $10,000 in two different funds. Part of it was invested at 2%, and the rest at 3%. She earned $240 in simple interest. How much did she invest at each rate? (*Hint:* Use the simple interest formula: $I = Prt$.)

Solve It

Answers to Problems on Weighing Quality and Quantity

The following are the answers (in bold) to the practice problems in this chapter.

1 Kathy's Kandies features a mixture of chocolate creams and chocolate-covered caramels that sells for $6 per pound. If creams sell for $4.50 per pound, and caramels sell for $7 per pound, how much of each type of candy should be in a one-pound mix? **.4 pounds of creams and .6 pounds of caramels**.

Let x = amount of chocolate creams in pounds. Then $1 - x$ = pounds of chocolate caramels. In a pound of the mixture, creams cost $4.50x$ and caramels $7(1 - x)$. Together, the mixture costs $6. So

$$4.5x + 7(1 - x) = 6$$
$$4.5x + 7 - 7x = 6$$
$$-2.5x + 7 = 6$$

Subtract 7 from each side and then divide by -2.5:

$$-2.5x = -1$$

$$x = \frac{-1}{-2.5} = .4 \text{ pounds of creams}$$

To get the amount of caramels, $1 - x = 1 - .4 = .6$ pounds of caramels. Checking this, the cost of creams is $4.50(.4) = 1.80$ dollars. The cost of caramels is $7(.6) = 4.20$ dollars. Adding these together, $1.80 + 4.20 = 6$ dollars.

2 Solardollars Coffee is trying new blends to attract more customers. The premium Colombian costs $10 per pound, and the regular blend costs $4 per pound. How much of each should the company use to make 100 pounds of a coffee blend that costs $5.50 per pound? **The company needs 25 pounds of Colombian and 75 pounds of regular blend**.

Let x = the pounds of Colombian coffee at $10 per pound. Then $100 - x$ = pounds of regular blend at $4 per pound. The cost of 100 pounds of the mixture blend is to cost $5.50(100) = \$550$. Use

$$10x + 4(100 - x) = 550$$
$$10x + 400 - 4x = 550$$
$$6x + 400 = 550$$

Subtract 400 from each side and then divide each side by 6:

$$6x = 150$$

$$x = \frac{150}{6} = 25$$

To check, multiply $10(25)$ and $4(75)$ to get $250 + 300 = 550$, as needed.

3 Peanuts cost $2 per pound, almonds cost $3.50 per pound, and cashews cost $6 per pound. How much of each should you use to create a mixture that costs $3.40 per pound, if you have to use twice as many peanuts as cashews? **⅜ pounds almonds, ⅖ pounds peanuts, ⅕ pounds cashews**.

Let x = the pounds of cashews at $6 per pound. Then $2x$ = pounds of peanuts at $2 per pound. The almonds are then $1 - (x + 2x) = 1 - 3x$ pounds at $3.50 per pound. Combine all this:

Chapter 20: Weighing In on Quality and Quantity Story Problems

$$6x + 2(2x) + 3.5(1 - 3x) = 3.4$$
$$6x + 4x + 3.5 - 10.5x = 3.4$$
$$-.5x + 3.5 = 3.4$$
$$-.5x = -.1$$
$$x = \frac{-.1}{-.5} = \frac{1}{5} \text{ pounds of cashews}$$

Use this amount for x and substitute in to get the other weights. You get 2(⅕) = ⅖ pounds of peanuts and 1 − 3(⅕) = 1 − ⅗ = ⅖ pounds of almonds.

4 A mixture of jellybeans is to contain twice as many red as yellow, three times as many green as yellow, and twice as many pink as red. How many of each color jellybean should be in a bag of 100 jellybeans? **10 yellow jellybeans, 20 red jellybeans, 30 green jellybeans, and 40 pink jellybeans.**

Let x = the number of yellow jellybeans. Then $2x$ = the number of red jellybeans, $3x$ = the number of green jellybeans, and $2(2x) = 4x$ the number of pink jellybeans. So the number of yellow jellybeans is 10:

$$x + 2x + 3x + 2(2x) = 100$$
$$10x = 100$$
$$x = \frac{100}{10} = 10$$

With 10 yellow jellybeans, you then know that the mixture has 20 red jellybeans, 30 green jellybeans, and 40 pink jellybeans. The 10 + 20 + 30 + 40 adds up to 100 jellybeans.

5 A *Very Berry Smoothie* calls for raspberries, strawberries, and yogurt. Raspberries cost $3 per cup, strawberries cost $1 per cup, and yogurt costs $0.50 per cup. The recipe calls for twice as much strawberries as raspberries. How many cups of strawberries are needed to make a gallon of this smoothie that costs $10.10? (*Hint:* 1 gallon = 16 cups) **1.2 cups**.

Let x = the cups of raspberries at $3 per cup. Then $2x$ = the cups of strawberries at $1 per cup. The remainder of the 16 cups of mixture are for yogurt, which comes out to be $16 - (x + 2x) = 16 - 3x$ cups of yogurt at $.50 per cup. The equation you need is

$$3(x) + 1(2x) + .5(16 - 3x) = 10.10$$
$$5x + 8 - 1.5x = 10.1$$
$$3.5x + 8 = 10.1$$
$$3.5x = 2.1$$
$$x = \frac{2.1}{3.5} = .6$$

The mixture has .6 cups of raspberries and 2(.6) = 1.2 cups of strawberries.

6 A supreme pizza contains five times as many ounces of cheese as mushrooms, twice as many ounces of peppers as mushrooms, twice as many ounces of onions as peppers, and four more ounces of sausage than mushrooms. If the topping on this pizza is to weigh 30 ounces, then how many ounces of each ingredient is used? **The mixture has 2 ounces of mushrooms, 10 ounces of cheese, 4 ounces of peppers, 8 ounces of onions, and 6 ounces of sausage**.

Let x = the ounces of mushrooms. Then $5x$ = the ounces of cheese, $2x$ = the ounces of peppers, $2(2x) = 4x$ the ounces of onions, and $x + 4$ ounces of sausage. The toppings weigh 30 ounces. So $x + 5x + 2x + 4x + x + 4 = 30$, which simplifies to $13x + 4 = 30$, $13x = 26$, $x = 2$. Fill the $x = 2$ into the original equations to determine each ingredient.

7 How many quarts of 25% solution do you need to add to 4 quarts of 40% solution to create a 31% solution? **6 quarts**.

Let x = the quarts of 25% solution needed. Add x quarts of 25% solution to 4 quarts of 40% solution to get $(x + 4)$ quarts of 31% solution. Write this $x(.25) + 4(.40) = (x + 4)(.31)$. Multiply through by 100 to get rid of the decimals: $25x + 4(40) = (x + 4)(31)$. Distribute the 31 and simplify on the left: $25x + 160 = 31x + 124$. Subtract $25x$ from each side and subtract 124 from each side to get $36 = 6x$. Divide by 6 to get $x = 6$. The answer is 6 quarts of 25% solution.

8 How many gallons of 5% solution have to be added to 2 gallons of 90% solution to create 15% solution? **15 gallons**.

Let x = the gallons of 5% solution. Then $(.05)x + (.90)(2) = (.15)(x + 2)$ for the $x + 2$ gallons. Multiply through by 100: $5x + (90)(2) = (15)(x + 2)$. Simplify each side: $5x + 180 = 15x + 30$. Subtract $5x$ and 30 from each side: $150 = 10x$ or $x = 15$ gallons of 5% solution.

9 How many quarts of pure antifreeze need to be added to 8 quarts of 30% antifreeze to bring it up to 50%? **3.2 quarts**.

Let x = the quarts of pure antifreeze to be added. Pure antifreeze is 100% antifreeze (100% = 1). So $1(x) + (.30)(8) = (.50)(x + 8)$. Simplify on the left and distribute on the right: $x + 2.4 = .5x + 4$. Subtract $.5x$ and 2.4 from each side: $.5x = 1.6$. Divide by .5 to get $x = 3.2$ quarts of pure antifreeze.

10 How many cups of chocolate syrup need to be added to 1 quart of milk to get a mixture that's 25% syrup? **1⅓ cups**.

Let x = the cups of chocolate syrup needed. I only use the best quality chocolate syrup, of course, so you know that the syrup is pure chocolate. 1 quart = 4 cups, and the milk has no chocolate syrup in it.

So

$$1(x) + 0(4) = (.25)(x + 4)$$
$$x = .25x + 1$$
$$.75x = 1$$
$$x = \frac{1}{.75} = \frac{100}{75} = \frac{4}{3} \text{ cups of chocolate syrup}$$

11 What concentration should the 4 quarts of salt water have so that when it's added to 5 quarts of 40% solution salt water the concentration goes down to 33⅓%? **25%**.

Let $x\%$ = the percent of the salt solution in the 4 quarts:

$$(x\%)(4) + (40\%)(5) = \left(33\frac{1}{3}\%\right)(9)$$
$$4x + 200 = \left(33\frac{1}{3}\right)(9)$$
$$4x + 200 = 300$$
$$4x = 100$$
$$x = 25$$

The 4 quarts must have a 25% salt solution.

12 What concentration and amount solution have to be added to 7 gallons of 60% alcohol to produce 16 gallons of 37½% alcohol solution? **9 gallons of 20% solution**.

To get 16 gallons, 9 gallons must be added to the 7 gallons. Let $x\%$ = the percent of alcohol in the 9 gallons:

$$(x\%)(9) + (60\%)(7) = \left(37\tfrac{1}{2}\%\right)(16)$$
$$9x + 420 = 600$$
$$9x = 180$$
$$x = \frac{180}{9} = 20$$

13 Carlos has twice as many quarters as nickels and has a total of $8.25. How many quarters does he have? **30 quarters.**

Let x = the number of nickels. Then $2x$ = the number of quarters. These coins total $8.25 or 825 cents. So, in cents,

$$5(x) + 25(2x) = 825$$
$$5x + 50x = 825$$
$$55x = 825$$
$$x = \frac{825}{55} = 15$$

Because Carlos has 15 nickels, he must then have 30 quarters.

14 Gregor has twice as many $10 bills as $20 bills, five times as many $1 bills as $10 bills, and half as many $5 bills as $1 bills. He has a total of $750. How many of each bill does he have? **10 $20 bills, 20 $10 bills, 100 $1 bills, and 50 $5 bills.**

Let x = the number of $20 bills. Then $2x$ = the number of $10 bills, $5(2x) = 10x$ = the number of $1 bills and $\tfrac{1}{2}(10x) = 5x$ = the number of $5 bills. The total in dollars is 750. So the equation should be $20(x) + 10(2x) + 1(10x) + 5(5x) = 750$. Simplify on the left to $75x = 750$ and $x = 10$. Gregor has 10 $20 bills, 20 $10 bills, 100 $1 bills, and 50 $5 bills.

15 Stella has 100 coins in nickels, dimes, and quarters. She has 18 more nickels than dimes and a total of $7.40. How many of each coin does she have? **40 dimes, 58 nickels, and 2 quarters.**

Let x = the number of dimes. Then $x + 18$ = the number of nickels, and the number of quarters is $100 - (x + (x + 18)) = 82 - 2x$. These coins total $7.40 or 740 cents:

$$10(x) + 5(x+18) + 25\left[100 - (x + (x+18))\right] = 740$$
$$10x + 5(x+18) + 25[82 - 2x] = 740$$
$$10x + 5x + 90 + 2{,}050 - 50x = 740$$
$$2{,}140 - 35x = 740$$
$$-35x = -1{,}400$$
$$x = \frac{-1{,}400}{-35} = 40$$

So Stella has 40 dimes, 58 nickels, and 2 quarters.

16 Betty invested $10,000 in two different funds. Part of it was invested at 2%, and the rest at 3%. She earned $240 in simple interest. How much did she invest at each rate? (**Hint:** Use the simple interest formula: $I = Prt$) **Betty has $6,000 invested at 2% and the other $4,000 invested at 3%.**

Let x = the amount invested at 2%. Then the amount invested at 3% is $10{,}000 - x$. Betty earns interest of 2% on x dollars and 3% on $10{,}000 - x$ dollars. The total interest is $240, so

$$(.02)(x) + (.03)(10{,}000 - x) = 240$$
$$.02x + 300 - .03x = 240$$
$$-.01x + 300 = 240$$
$$-.01x = -60$$
$$x = \frac{-60}{-.01} = 6{,}000$$

Part V
The Part of Tens

In This Part...

Did you ever think about why numbers are in base ten? Why is everything ten or one hundred or one tenth? Why aren't they in base six? (What seems to leap out at me is right at the end of my hands. If everyone had been born with 12 fingers, then we'd be using base 12!)

So, doesn't it seem logical that this last part has lists that have ten items in them? These three chapters provide quick lists of ten items in *For Dummies* fashion. Specifically they give you ten basic graphing facts, ten common pitfalls to avoid, and ten quick tips to make your algebra problem solving a little easier.

Chapter 21

Ten (or So) Things to Know about Graphing

In This Chapter
- Plotting points and lines
- Finding quadrants, midpoints, and distances
- Determining slopes of lines
- Intercepting and intersecting
- Working with parabolas

Graphs are as important to algebra as pictures are to books and magazines. A graph can represent data that you've collected, or it can represent a pattern or model of an occurrence. A graph illustrates what you're trying to demonstrate.

The standard system for graphing in algebra is to use the *Cartesian coordinate system,* where points are represented by ordered pairs of numbers, and connected points can be lines, curves, or disjointed pieces of graphs.

This chapter outlines the basic concepts and strategies you need to know about graphing. If midpoints and intersections give you trouble, this chapter can help you sort it all out.

Thickening the Plot with Points

Graphing on the Cartesian coordinate system involves drawing two perpendicular axes, the x-axis horizontal and the y-axis vertical. The Cartesian coordinate system identifies a point by an ordered pair, (x, y). Remember that the order in which the coordinates are written matters. The first coordinate, the x, represents how far to the left or right the point is from the *origin,* or where the axes intersect. A positive x is to the right; a negative x is to the left. The second coordinate, the y, represents how far up or down from the origin the point is.

The Cartesian coordinates designate where a point is in reference to the two perpendicular axes. To the right and up is positive, to the left and down is negative. Any point that lies on one of the axes has a 0 for one of the coordinates. The coordinates for the *origin,* the intersection of the axes, are $(0, 0)$.

260 Part V: The Part of Tens

Q. Use the following figure to graph the points (2, 6), (8, 0), (5, –3), (0, –7), (–4, –1), and (–3, 4).

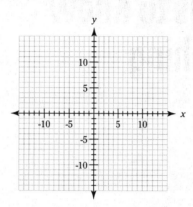

A. Notice that the points that lie on an axis have a 0 in their coordinate.

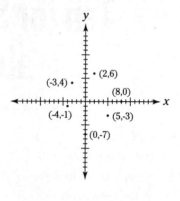

Sectioning Off by Quadrants

Another designation in graphing uses the *quadrant* that a point lies in. The quadrants are referred to in many applications because of the common characteristics of points that lie in the same quadrant. The quadrants are numbered one through four, usually with Roman numerals. Check out Figure 21-1 to see how the quadrants are identified.

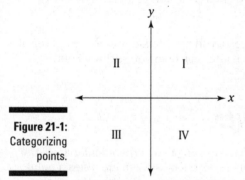

Figure 21-1: Categorizing points.

Chapter 21: Ten (or So) Things to Know about Graphing   261

Q. Referring to Figure 21-1, describe which coordinates are positive or negative in the different quadrants.

A. In quadrant I, both the *x* and *y* coordinates are positive numbers. In quadrant II, the *x* coordinate is negative, and the *y* coordinate is positive. In quadrant III, both the *x* and *y* coordinates are negative. In quadrant IV, the *x* coordinate is positive, and the *y* coordinate is negative.

Plotting Points for Lines

One of the most basic graphs you can do using the coordinate system is the graph of a straight line. You may remember, from geometry, that it only takes two points to determine a line. When graphing lines using points, though, plot three points to be sure that you've found correct points and put them in the correct positions. You can think of the third point as a sort of a check. The third point can be anywhere, but try to spread out the three points and not have them clumped together. If the three points aren't in a straight line, then at least one of them is wrong.

Q. Use the following figure to graph the line represented by the equation $2x + 3y = 10$.

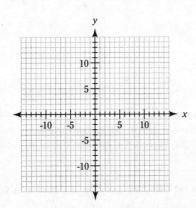

A. Find three sets of coordinates that satisfy the equation. **Three points that work for this line are (5, 0), (2, 2), and (–1, 4)**. Plot the three points, and then draw a line through them.

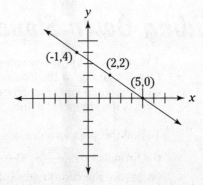

Graphing Lines with Intercepts

An *intercept* is a point that a figure shares with one of the axes. Vertical lines have one intercept — it's a point somewhere on the *x*-axis. Horizontal lines have one intercept — and that's a point on the *y*-axis. All other lines cross both of the axes and have two intercepts.

Intercepts are easy to find when you have the equation of a line. To find the *x*-intercept, you let *y* be 0, and solve for *x*. To find the *y*-intercept, you let *x* be 0, and solve for *y*.

Q. Find the intercepts of the line $9x - 4y = 18$.

A. $(2,0)$ and $\left(0, -\frac{9}{2}\right)$. First, to find the *x* intercept, set $y = 0$ in the equation of the line to get $9x = 18$. Solving that, $x = 2$, and the intercept is $(2, 0)$. Next, for the *y* intercept, set $x = 0$ to get $-4y = 18$. Solving for *y*, $y = \frac{18}{-4} = -\frac{9}{2}$. Intercepts make it nice for graphing the line, too. Use the two intercepts to graph the line, and then you can check with one more point. For instance in the following figure, I graph the line and check to see if the point $(\frac{2}{9}, -4)$ is on the line.

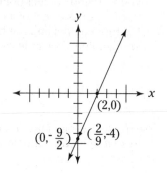

Sliding Down Slopes

The *slope* of a line is a number that describes the steepness of the line and whether it's rising or falling, as the line moves from left to right in a graph. To determine how steep a line is, when you're given its slope, the general rule is that the farther the number is from 0, the steeper the line.

To find the slope of a line, you can use two points on the graph of the line and apply the formula $m = \frac{y_2 - y_1}{x_2 - x_1}$. The *m* is the traditional symbol for slope; the (x_1, y_1) and (x_2, y_2) are the coordinates of two points on the line. The point you choose to go first in the formula doesn't really matter. Just be sure to keep the order the same for the *y*'s and *x*'s — you can't mix and match.

A *horizontal* line has a slope of 0, and a vertical line has no slope. To help you remember, picture the sun coming up on the *horizon* — that 0 is just peeking out at you.

Q. Find the slope of the line that goes through the two points (–3, 4) and (1, –8) and use the following figure to graph it.

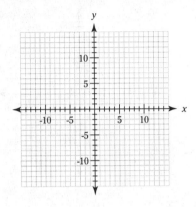

A. –3. To find the slope, use $m = \frac{4-(-8)}{-3-1} = \frac{12}{-4} = -3$. The following figure shows a graph of that line. It's fairly steep — any slope greater than 1 or less than –1 is steep. The negative part indicates that it's falling as you go from left to right. Think of the slope as being $\frac{-3}{1}$. The bottom number is the *change in x*, and the top is the *change in y*. The way you read it is: *For every 1 unit you move to the right parallel to the x-axis, you drop down 3 units parallel to the y-axis.*

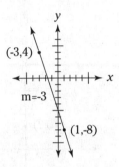

Graphing with the Slope-Intercept Form

Equations of lines can take many forms, but one of the most useful is called the *slope-intercept form*. When you look at this form on a graph, you slope the line and its *y*-intercept.

 The slope intercept form is $y = mx + b$. The *m* represents the slope of the line. The *b* is the *y*-coordinate of the intercept where the line crosses the *y*-axis. A line with the equation $y = -3x + 2$ has a slope of –3 and a *y*-intercept of (0, 2).

Having the equation of a line in this form makes graphing the line an easy chore. Follow these steps:

1. **Plot the *y*-intercept on the *y*-axis, and then count off the slope.**

2. **Write the slope as a fraction.**

   Using the equation $y = -3x + 2$, the fraction would be $\frac{-3}{1}$

3. **If the slope is negative, put the negative part in the numerator.**

   The slope has the change in *y* in the numerator and the change in *x* in the denominator.

4. **Starting with the *y*-intercept, count the amount of the change in *x* to the right of the intercept, and then count up or down (depending on whether it's positive or negative) from that point.**

   Wherever you end up is another point on the line.

5. **Mark that point and draw a line through the two points.**

Part V: The Part of Tens

Q. Graph $y = -3x + 2$ using the method in the previous steps.

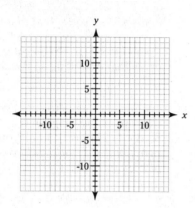

Q. Graph $y = \frac{2}{5}x - 1$.

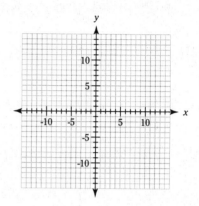

A.

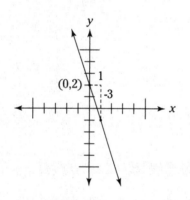

A.

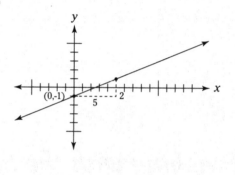

Changing to the Slope-Intercept Form

Graphing lines using the slope-intercept form is a piece of cake. But what if the equation isn't in that form? Are you stuck with substituting in values and finding coordinates of points that work? Not necessarily. Changing the form of the equation using algebraic manipulations — and then graphing using the new form — is often easier.

To change the equation of a line to the slope-intercept form, $y = mx + b$, first isolate the term with y in it on one side of the equation, and then divide each side by any coefficient of y. You can rearrange the terms so the x term, with the slope multiplier, comes first.

Chapter 21: Ten (or So) Things to Know about Graphing **265**

Q. Change the equation $3x - 4y = 8$ to the slope-intercept form.

A. $y = \frac{3}{4}x - 2$. First, subtract $3x$ from each side: $-4y = -3x + 8$. Then divide each term by -4:

$$\frac{-4y}{-4} = \frac{-3x}{-4} + \frac{8}{-4}$$
$$y = \frac{3}{4}x - 2$$

This line has a slope of ¾ and a *y*-intercept at $(0, -2)$.

Q. How do you know when a slope is too steep?

A. **When you land on your derrière**. Hey, skiing is a lot like algebra. They're both dangerous if you don't prepare yourself for what's ahead. Make sure you take the time to practice your algebra skills so you don't end up on your backside.

Lining Up Parallel and Perpendicular Lines

When two lines are *parallel* to one another, they never touch, and their slopes are exactly the same. When two lines are *perpendicular* to one another, they cross in exactly one place, and their slopes are related. If two perpendicular lines aren't vertical and horizontal (parallel to the two axes), then one has a positive slope, one has a negative slope, and the number values of the slopes are reciprocals. (The *reciprocal* of a number is its flip, what you get when you put the number in the bottom of a fraction with a 1 in the top.) In other words, the slopes are *negative reciprocals* of one another.

Q. Determine how these lines are related: $y = \frac{6}{2}x + 1$, $y = \frac{15}{5}x - 11$, and $y = 3x + 4$

A. **They're parallel**. They're written in slope-intercept form and have the same slope. The *y*-intercepts are the only differences between these two lines.

Q. Determine how these lines are related: $y = -\frac{4}{9}x + 3$ and $y = \frac{9}{4}x - 19$

A. **They're perpendicular to one another**. Their slopes are negative reciprocals of one another. It doesn't matter what the *y*-intercepts are. They can be different or the same.

Finding Distances Between Points

A segment can be drawn between two points on the coordinate axes. You can determine the distance between those two points by using a formula that actually incorporates the Pythagorean theorem — it finds the length of a hypotenuse of a right triangle. (Check Chapter 17 for more practice with the Pythagorean theorem.) If you want to find the distance between the two points (x_1, y_1) and (x_2, y_2), use the formula $d = \sqrt{(x_1 - x_2)^2 + (y_1 - y_2)^2}$.

Q. Find the distance between the points (–8, 2) and (4, 7).

A. **13 units**. Use the distance formula and plug in the coordinates of the points:

$$d = \sqrt{(-8 - 4)^2 + (2 - 7)^2}$$
$$= \sqrt{(-12)^2 + (-5)^2}$$
$$= \sqrt{144 + 25}$$
$$= \sqrt{169}$$
$$= 13$$

Of course, not all the distances come out nicely with a perfect square under the radical. When it isn't a perfect square, either simplify the expression or estimate the answer, as in Chapter 5. When the two points are along the same horizontal or vertical line, then one of the terms under the radical becomes a 0, and you get a perfect square under the radical.

Q. Find the distance between the points (4, –3) and (4, 11).

A. **14 units**. Using the distance formula, you get

$$d = \sqrt{(4 - 4)^2 + (-3 - 11)^2}$$
$$= \sqrt{0^2 + (-14)^2}$$
$$= \sqrt{(-14)^2}$$
$$= 14$$

Actually, though, the distance formula is unnecessary in these cases. If you notice that the two *x*-coordinates of the points are the same, just find the absolute value of the distance between the *y*-coordinates, and vice versa.

Plotting Parabolas

A *parabola* is a sort of U-shaped curve. It's one of the conic sections (the other conics are a circle, hyperbola, and ellipse). The equations and graphs of parabolas are used to describe all sorts of natural phenomena. For instance, headlight reflectors are formed from parabolic shapes.

You can tell all sorts of interesting things about the particular parabolic curve from the equation of a parabola. In this discussion, I stick just to parabolas that open upward and downward. The general equation for all these parabolas is $y = ax^2 + bx + c$, where *a* isn't 0.

You can identify several characteristics of the parabola with this general equation, including:

- The coordinates of the *vertex*
- The equation of the *axis of symmetry*
- Whether it opens upward or downward
- What some other points on the parabola are

To find these characteristics of a parabola, follow these steps:

1. **You can find the *x*-coordinate of the vertex by using the coefficients *a* and *b* in the formula $x = \frac{-b}{2a}$.**

   The *vertex* is the maximum or minimum point of the parabola — depending on whether it's opening upward or downward.

2. **You can then find the *y*-coordinate by substituting this *x* value back into the equation for the parabola.**

3. **To find the axis of symmetry, write the equation as *x* equal to the *x*-coordinate of the vertex.**

   The *axis of symmetry* is the line through the vertex that divides the parabola into two equal halves — mirror images of one another on either side of the axis. The parabola opens upward if the coefficient *a* is positive and downward if it's negative.

4. **To graph the parabola, you pick another *x* value, substitute into the equation to get the *y*-value, and then graph the point with those *x* and *y* values.**

   The *pair* of that random point is on the other side of the axis of symmetry. These points help you to sketch the graph of the parabola.

Look at this figure, which is the graph of the parabola $y = -2x^2 + 4x + 7$, and you can determine the following information about the parabola:

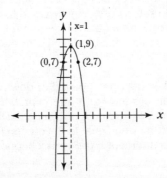

- Vertex of the parabola: (1, 9)
- Axis of symmetry: $x = 1$
- Opens downward
- Another point: (0, 7) and its pair: (2, 7)

Q. Find the vertex, axis of symmetry, and two other points on the parabola $y = -3x^2 - 18x - 20$ and sketch its graph.

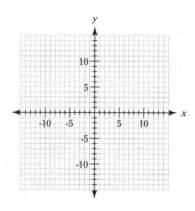

A. Vertex: $(-3, 7)$; axis of symmetry: $x = -3$; two other points (these may vary): $(0, -20)$ and $(-4, 4)$.

To find the vertex, $x = \dfrac{-(-18)}{2(-3)} = \dfrac{18}{-6} = -3$ and $y = -3(-3^2) - 18(-3) - 20 = -27 + 54 - 20 = 7$.

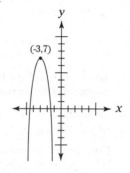

Taking on Intercepts of Parabolas

Intercepts of graphs are where the graphs cross or touch the axes. The graph of the parabola $y = -x^2 - 8x + 20$ has three intercepts: a y-intercept of $(0, 20)$ and two x-intercepts of $(-10, 0)$ and $(2, 0)$. The intercepts not only help when you're graphing the parabola, but they're also useful when the equation is being used for a particular application. The y-intercept can represent a beginning or initial value, and the x-intercepts can tell you when you switch from positive to negative values or vice versa.

To find the y-intercept, you just let x be equal to 0, substitute that into the equation, and solve for y. To find the x-intercepts, let y be equal to 0. When you substitute that into the equation, you end up with a quadratic equation. Try factoring, first. If that fails, go to the quadratic formula. (See Chapter 13 for more practice on solving quadratic equations.)

When dealing with parabolas that open upward or downward, you always have a y-intercept, but you don't always have an x-intercept. If the parabola opens upward and has its vertex above the x-axis, then it doesn't cross the x-axis. When you solve the quadratic equation after letting y be equal to 0, you don't find any real solutions. Also, the parabola has only one x-intercept if the vertex happens to be on the x-axis.

Q. Find the intercepts of $y = -x^2 - 8x + 20$.

A. $(0, 20)$, $(-10, 0)$, and $(2, 0)$. First, solve for the y-intercept by letting $x = 0$: $y = -0^2 - 8(0) + 20 = 20$. When $x = 0$, you get a y-intercept of 20. Write it as $(0, 20)$. Now solve for the x-intercepts by letting y be equal to 0: $0 = -x^2 - 8x + 20$. If you factor out the -1 first, then you can deal with it easier: $0 = -(x^2 + 8x - 20) = -(x + 10)(x - 2)$. Now, setting each of the factors equal to 0, you get $x = -10$ or $x = 2$. That's what you get when you let $y = 0$.

Graphing with Transformations

You can graph the curves and lines associated with different equations in many different ways. Finding the vertex or intercepts of a parabola, the intercepts of a line, or any of the other types of determining points of a graph are all helpful when sketching the graph of the figure.

In many instances, graphs are alike in shape and steepness, but they're just in different positions on the axes. A change in position happens when a curve undergoes a *transformation* called a *translation* or slide. It slides to the left, to the right, up, or down. You can take the basic parabola, for instance, and slide it around using the following rules. The *C* represents some positive number:

$y = x^2 + C$: raises the parabola by C units.

$y = x^2 - C$: lowers the parabola by C units.

$y = (x + C)^2$: slides the parabola left by C units.

$y = (x - C)^2$: slides the parabola right by C units.

A graph can have its steepness changed by multiplying it by a number. If the basic function or operation is multiplied by a positive number greater than 1, then it becomes steeper. If it's multiplied by a positive number smaller than 1, it becomes flatter. Multiplying a basic function by a negative number results in a flip or reflection over a horizontal line. Use these rules when using the parabola:

$y = kx^2$: Parabola becomes steeper when k is positive and greater than 1.

$y = kx^2$: Parabola becomes flatter when k is positive and smaller than 1.

$y = -x^2$: Parabola flips over a horizontal line.

Q. Use the following eight graphs to graph $y = x^2 + 1$, $y = x^2 - 1$, $y = (x + 1)^2$, $y = (x - 1)^2$, $y = 3x^2$, $y = \frac{1}{3}x^2$, $y = -2x^2$, $y = (x - 1)^2 + 1$.

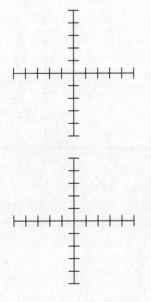
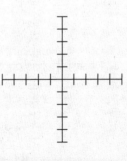

270 Part V: The Part of Tens

A.

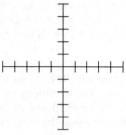
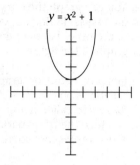
$y = x^2 + 1$

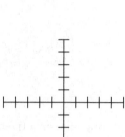
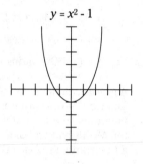
$y = x^2 - 1$

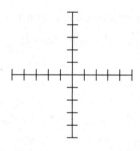
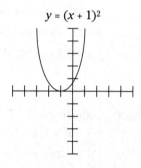
$y = (x + 1)^2$

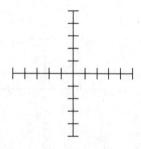
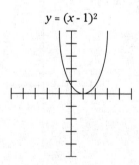
$y = (x - 1)^2$

Chapter 21: Ten (or So) Things to Know about Graphing 271

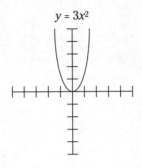

$y = 3x^2$

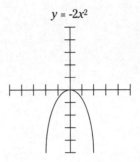

$y = -2x^2$

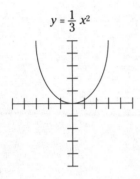

$y = \frac{1}{3} x^2$

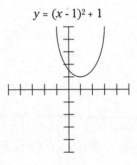

$y = (x - 1)^2 + 1$

Chapter 22
Ten Common Pitfalls and How to Avoid 'Em

In This Chapter
- Empowering with powers
- Looking at fraction errors
- Making sense of inequality senses

Most of this book involves ways to correctly do algebra procedures. When working these problems, take the positive approach. But sometimes I may need to point out the common errors or pitfalls of working with algebra. Many of the pitfalls involve the negative sign, a fraction, or an exponent. Mistakes happen in algebra because many people think what they're doing works. However, that's not always the case.

This chapter includes the pitfalls that happen often, to all sorts of people. Read this chapter and make yourself aware so you don't end up tripping over them.

Squaring a Power

The rule for raising a power to a power is that you multiply the exponents: $(x^a)^2 = x^{2a}$

The correct way to solve is $(x^4)^2 = x^8$ or $(y^{6z})^2 = y^{12z}$.

A common error is to square the power itself rather than multiplying it by 2. For example, $(x^5)^2 \neq x^{25}$, $(x^5)^2 = x^{10}$ or $(y^{3a})^2 \neq y^{9a^2}$, $(y^{3a})^2 = y^{6a}$.

Refer to Chapter 4, if you need more information on working with exponents.

Squaring a Binomial

A *binomial* is two terms separated by addition or subtraction. Squaring a binomial involves more than just squaring each term — the terms have to interact with one another. The pattern for squaring a binomial is $(a + b)^2 = a^2 + 2ab + b^2$.

When you square a binomial, remember that you're multiplying it times itself: $(a + b)^2 = (a + b)(a + b)$. You have the products of the first terms, the outer terms, the inner terms, and the last terms. The right way to solve is $(x + 5)^2 = x^2 + 10x + 25$ or $(3y - 2z)^2 = 9y^2 - 12yz + 4z^2$.

A common error is to just square the first and last term of the binomial and forget about the middle term. For example, $(z+3)^2 \neq z^2 + 9$, $(z+3)^2 = z^2 + 6z + 9$ or

$$(3t^4 - 1)^2 \neq 9t^8 + 1$$
$$(3t^4 - 1)^2 = 9t^8 - 6t^4 + 1$$

You can find more on squaring binomials in Chapter 7, if you want some more details and practice.

Ordering Around Operations

The order of operations dictates that when you have several choices of operations to perform, you do powers and roots first, then multiplication and division, and lastly addition and subtraction.

A common error occurs with an expression such as -3^2. This expression involves a power and a subtraction. (You read the minus sign, negative sign, or subtraction differently but they all fall in the same category). In an expression like this, do the power first and then negate the answer: $-3^2 = -(3^2) = -(9) = -9$ or $-5^4 = -(5^4) = -(625) = -625$.

If you want to perform the power on the entire number, negative sign and all, you need to have a parenthesis around it: $(-3)^2 = (-3)(-3) = +9$.

Check out Chapter 6 for more on the order of operations.

Becoming Radical

The square root of a product is the product of the roots, and the square root of a quotient is the quotient of the roots, but you can't just do the operations to the separate terms when you have a sum or difference. With a product or quotient, everything is bound together by those operations into one, compact term. Sums and differences separate the expression into more than one term. In this following example, I start out with two terms under the radical, but change it to one term by factoring out the 4. Now it's a product of 4 times $2 + x$. I separate those two factors, and then compute the square root of each factor:

$$\sqrt{8 + 4x} = \sqrt{4(2+x)} = \sqrt{4}\sqrt{2+x} = 2\sqrt{2+x}$$

In this example, the square root of a product is equal to the product of the roots. The next two examples are the errors that occur:

$$\sqrt{x^2 + 9} \neq x + 3$$
$$\sqrt{9 + 16} = \sqrt{25} = 5$$
$$\sqrt{9 + 16} = \sqrt{3^2 + 4^2} \neq 3 + 4$$

A common error is to try to take the roots of the two terms under the radical. Refer to Chapter 5 for more on working with radicals.

Distributing a Negative

Distributing a number by multiplying it times every term within a parenthesis is a fairly straightforward process. The distributive property of multiplication over addition is $a(b + c) = ab + ac$. The pitfall or problem comes in when distributing a negative (the same as -1) over terms. Often, people distribute it over the first term and not the rest.

This is the correct way: $-(x^2 + 2x - 3) = (-1)x^2 + (-1)2x - (-1)(3) = -x^2 - 2x + 3$

This is the common error: $-(z^4 - 3z^3 - 3z^2 - 9) \neq -z^4 - 3z^3 - 3z^2 - 9$

Remember to distribute the negative over every term. In the previous example, the last three terms all should have changed to positive. Chapter 2 has even more on the distribution rule.

Fracturing Fractions

A fraction line is like a grouping symbol. Everything in the *denominator* (bottom) divides into everything in the *numerator* (top).

The correct way to split up a fraction with several terms in the numerator is to write each term over the entire denominator: $\dfrac{x^2 - 2}{x + 7} = \dfrac{x^2}{x + 7} - \dfrac{2}{x + 7}$

So the following is a big no-no:

$$\dfrac{4+8}{2+4} \neq \dfrac{4}{2} + \dfrac{8}{4}$$
$$\dfrac{12}{6} \neq 2 + 2$$

Chapter 3 covers fractions completely.

Reducing Fractions

A fraction is reduced properly when the numerator and denominator are divided by the same number, which means the *entire* numerator and denominator, not just a term or two in each.

Here's the correct way to reduce fractions in algebra.

$\dfrac{6 + 3x}{6} = \dfrac{3(2+x)}{6} = \dfrac{\cancel{3}^{1}(2+x)}{\cancel{6}_{2}} = \dfrac{2+x}{2}$. The 3 multiplies the entire numerator and denominator.

$\dfrac{x^2 - 4}{x - 2} = \dfrac{(x-2)(x+2)}{x-2} = \dfrac{\cancel{(x-2)}(x+2)}{\cancel{x-2}} = \dfrac{x+2}{1} = x + 2$. Factor the numerator first.

The following are two very common errors in reducing fractions: $\dfrac{6+x}{6} \neq \dfrac{\cancel{6}+x}{\cancel{6}}$ or $\dfrac{(x+2)^8 - 9}{(x+2)^3} \neq \dfrac{(x+2)^{\cancel{8}}-9}{(x+2)^{\cancel{3}}}$

Refer to Chapter 3 if you want more practice on working with fractions.

Using Negative Exponents

The general rule for negative exponents is $\frac{1}{x^n} = x^{-n}$. Negative exponents are very handy when combining terms with the same base. A common pitfall is to forget about number multipliers when applying this rule:

$$\frac{1}{2x^3} \neq 2x^{-3}$$

$$\frac{1}{2x^3} = 2^{-1}x^{-3} \text{ or } \frac{1}{2}x^{-3}$$

The exponent on the 2 is assumed to be a 1.

Check out Chapter 4 for more on negative exponents.

Determining Which Is Smaller

Comparing 13 with 30 is easy. You can quickly tell which number is larger and which is smaller. However, negative numbers can cause the biggest hang-up — especially when fractions are involved. When comparing two negative numbers, the one that's farther from 0 is the smaller. Another way to figure out which is smaller is to look at absolute value. The one with the greater absolute value is smaller. Check out these comparisons.

$-7 < -3$

$-4\frac{1}{2} < -4\frac{1}{3}$. The larger the denominator, the smaller the number.

$-6.0001 < -6.00001$. The decimal with hundred-thousandths is closer to 0.

Chapter 1 covers comparing numbers on the number line.

Reversing the Sense

Because reversing the sense in an inequality isn't used that often, people frequently overlook it. The inequality $3 < 4$ is a true statement. When do you *reverse the sense* (switch the direction of the inequality sign)? You reverse it when you multiply or divide each side by a negative number. The other operations performed on inequalities don't have that kind of requirement.

$-5 < 7$

Now, multiplying each side by -1:

$(-1)(-5) > (-1)(7)$
$5 > -7$

Can you see that, if the inequality hadn't been turned around, the statement wouldn't have been true? Chapter 16 describes how to deal with inequalities.

Chapter 23

Ten Quick Tips to Make Algebra a Breeze

In This Chapter

- Working efficiently with fractions
- Zeroing in on the solution
- Devising divisibility rules
- Multiplying through to make it easier

Algebra is a lot like filling out tax forms. No matter what you do, you need to follow the rules. And, buried in those lists of rules are some maneuvers, procedures, and quick tricks that help to make the process better. What is *better*? Better may include eliminating fractions or decimals. Better may mean simplifying the set-up or changing the form to make the equation available to combine with other expressions. And, sometimes, better is just in the eyes of the beholder.

This chapter offers you ten quick tips to help make your experience with algebra better (and a little easier).

Flipping Proportions

A *proportion* is an equation in which one fraction is set equal to another fraction. One property of a proportion is that *flipping* the entire equation doesn't change its truth. The rule is if $\frac{a}{b} = \frac{c}{d}$, then $\frac{b}{a} = \frac{d}{c}$.

Fill in some numbers to convince yourself: $\frac{1}{2} = \frac{3}{6}$ is true, and so is $\frac{2}{1} = \frac{6}{3}$.

Now look at a situation where flipping makes the statement easier to solve. $\frac{8}{x+1} = \frac{1}{3}$ is a proportion. Now flip it to get the variable in the numerator: $\frac{x+1}{8} = \frac{3}{1}$. Multiplying each side by 8 gets rid of the denominator $x + 1 = 24$, and subtracting 1 from each side isolates the variable on the left and solves the equation: $x = 23$

Multiplying Through to Get Rid of Fractions

As much fun as fractions are — we couldn't do without them — they're sometimes a nuisance when you're trying to solve for the value of a variable. A quick trick is to multiply both sides of the equation by the common denominator. This common denominator is also called a common multiple — all the denominators divide it evenly. See the following example where I multiply each fraction by 12. See how easy it becomes.

$$\frac{x}{6} + \frac{2x-1}{3} = \frac{1}{4}$$

$$\left(\frac{x}{6}\right)(12) + \left(\frac{2x-1}{3}\right)(12) = \left(\frac{1}{4}\right)(12)$$

Each term has a fraction that reduces to have a denominator of 1.

$$2x + (2x-1)(4) = 3$$
$$2x + 8x - 4 = 3$$
$$10x - 4 = 3$$
$$10x = 7, \ x = \frac{7}{10}$$

Zooming In on the Zero

When you're solving an equation that involves a fraction set equal to 0, you only need to consider the fraction's numerator.

You have to completely reduce the fraction; you can't have any common factors in the numerator and denominator. The reason you only have to consider the numerator is that for a fraction to equal 0, the numerator must equal 0, and the denominator can be anything but 0.

For instance, $\frac{(x-2)(x+11)}{x^4(x+7)(x-9)(x+5)}$ is equal to 0 only when the numerator $(x-2)(x+11)$ equals 0. This equation is true when $x = 2$ or $x = -11$. These are the only two solutions. You can forget about the denominator, as long as you don't use a number that makes it equal to 0.

Finding a Common Denominator

When adding or subtracting fractions, you need to write all the fractions with the same common denominator.

I already provide a quick tip for multiplying through by a common denominator in this chapter. (See "Multiplying Through to Get Rid of Fractions" earlier in this chapter.) Most of the time you can easily spot the common denominator. For example, when the denominators are 2, 3, and 6, you can see that they all divide 6 evenly — 6 is the common denominator. Even numbers like 3 and 7 have a common denominator that's easy to spot — you just multiply the two numbers together.

For instance, if you want to add $\frac{x}{24} + \frac{3x-5}{60}$, you want to find a common denominator that isn't any bigger than necessary. Multiply the 24×60 to get 1,440. But 24 and 60 are both divisible by 12, so divide the 1,440 by 12 to get 120, which is the least common denominator.

Dividing by 3 or 9

Reducing fractions makes them easier to work with. The numbers are smaller in reduced fractions, and the common factors have been eliminated to make them simpler. What if you're trying to reduce a fraction that has numbers in it that are beyond those multiplication tables that you mastered years ago? For instance, do you know the common factor for $\frac{10071}{30006}$? You can divide both the numerator and denominator by 9. Not obvious to you? Keep reading.

A number is divisible by 3 if the sum of its digits is divisible by 3. So the number 1,047 is divisible by 3, because the sum of its digits is 12, which is divisible by 3. Also, a number is divisible by 9 if the sum of its digits is divisible by 9. So the number 10,071 is divisible by 9, because its digits add up to 9. The number 123,453 is divisible by 9, because the sum of its digits is 18, which is divisible by 9. And look at the number 18; the sum of its digits is 9. The number 30,006 is divisible by 9, because the digits add up to 9.

Now, looking at $\frac{10071}{30006} = \frac{1119}{3334}$, you see that I divided both numerator and denominator by 9 to reduce the fraction. The reduced fraction doesn't have any common factors in the numerator and denominator. Even though the new numerator is still divisible by 3, the denominator isn't. They don't have any more common factors, so the fraction is reduced.

This business of adding up the digits only works for 3 and 9 and other powers of 3, so don't try it for other numbers.

Dividing by 2, 4, or 8

The rules for seeing if a number is divisible by 2, 4, or 8 are quite different from the rules for 3 and 9.

No matter how big the number is, you only have to look at the last digit to see if a number is divisible by 2. You only have to look at the number formed by the last two digits to see if a number is divisible by 4, and at the last three digits for 8.

If a number ends in 0, 2, 4, 6, or 8 (it's an even number), the number is divisible by 2. The number 113,579,714 is divisible by 2, even though all the digits except the last are odd.

If the number formed by the last *two* digits of a number is divisible by 4, then the entire number is divisible by 4. The number 5,783,916 is divisible by 4, because the number formed by the last two digits, 16, is divisible by 4.

If the number formed by the last *three* digits of a number is divisible by 8, then the whole number is divisible by 8. The number 43,512,619,848 is divisible by 8, because the number formed by the last three digits, 848, is divisible by 8. You may run into some three-digit numbers that aren't so obviously divisible by 8. In that case, just do the long division on the last three digits. Dividing the three-digit number is still quicker than dividing the entire number.

Commuting Back and Forth

The *commutative law of addition and multiplication* says that you can add or multiply numbers in any order, and you'll get the same answer. This rule is especially useful and helpful when you couple it with the associative rule that allows you to regroup or reassociate numbers to make computations easier. For instance, look at how I can rearrange the numbers to my advantage for doing computations.

$$9 + 27 + 11 + 3 = 9 + 11 + 27 + 3 = (9 + 11) + (27 + 3) = 20 + 30 = 50$$

$$9(8)(25)\left(\frac{5}{8}\right)\left(\frac{7}{9}\right)\left(\frac{1}{25}\right) = \left[(8)\left(\frac{5}{8}\right)\right]\left[(9)\left(\frac{7}{9}\right)\right]\left[(25)\left(\frac{1}{25}\right)\right] = (5)(7)(1) = 35$$

Lining Up with Symmetry

The *symmetric property* says that if $a = b$, then $b = a$ is also true. You may think that concept seems obvious, but it's useful when solving equations. You want to take advantage of this property to get everything lined up.

For instance, if you're solving the equation $9x - 2 = 11x + 10$ and subtract $9x$ and 10 from each side, you get $-12 = 2x$. Dividing each side by 2, $-6 = x$. Many people prefer to write this as $x = -6$, because it reads more like you'd give the answer.

Now you know that you don't have to hesitate. Switch the equation around, if you want, to make it more readable and pleasing to your eye.

Making Radicals Less Rad, Baby

Radicals are symbols that indicate an operation. Whatever is inside the radical has the root operation performed on it. Many times, though, writing the radical expressions with exponents instead of radicals is more convenient. Doing so is especially helpful when you have to combine several terms or factors with the same variable under different radicals.

For instance, you may not see how to deal with $\left(\sqrt{y}\right)\left(\sqrt[3]{y}\right)\left(\sqrt[5]{y^2}\right)$. But you can multiply them all together, because the base y is the same in each.

First, the general rule for changing from radical form to exponential form is $\sqrt[b]{x^a} = x^{a/b}$.

Take the radical expression $\left(\sqrt{y}\right)\left(\sqrt[3]{y}\right)\left(\sqrt[5]{y^2}\right)$ and change it to $y^{1/2} \, y^{1/3} \, y^{2/5}$.

Now you can multiply the factors together by adding the exponents. Of course, you need to change the fractions so they have a common denominator, but the result is worth the trouble: $y^{1/2} \, y^{1/3} \, y^{2/5} = y^{15/30} \, y^{10/30} \, y^{12/30} = y^{37/30} = y^{17/30}$.

Eying Up the Polynomial Function

The graph of the polynomial $y = a_n x^n + a_{n-1} x^{n-1} + a_{n-2} x^{n-2} + \ldots + a_1 x^1 + a_0$ is a smooth curve that can touch or cross the *x*-axis, switch back and forth with *turns*, and rise or fall to infinitely high or low values. The way to tell the *maximum* number of *x*-intercepts and the *maximum* number of turning points that a polynomial can have is to look at its highest power, n.

If the highest power in the polynomial function is n, then the function has a *maximum* of n *x*-intercepts and a *maximum* of $n - 1$ turning points.

Look at the polynomial $y = 4x^5 - 5x^4 + 2x^3 - 3x^2 + x + 11$. Because the highest power is 5, the polynomial has at most five *x*-intercepts and four turning points. I keep harping on *maximum*, because the polynomial may not have that many — it could be any number less than that maximum number — but it'll never have more than that maximum number.

Index

• A •

absolute value
  grouping and, 17
  number lines, 10–11
  signed numbers, adding, 9–11
absolute value equations
  answers to problems, 184–188
  description, 177
  solving for, 182–183
absolute value inequalities, 195–197
acute angles, story problems and geometry, 224
addition
  expressions, 67
  fractions, 32–33
  signed numbers, 9–11
addition/subtraction property, linear
    equations, 131–132
age story problems, 235–237
algebraic properties
  answers to problems, 23–24
  associative rule, 20–21
  commutative property, 21–22
  distributive property, 19–20
  grouping symbols, 17–18
area, story problems and, 222–223
area formulas, 210–211
associative rule, grouping and, 20–21
axis of symmetry, parabola graphing, 267–268

• B •

binomials
  cubing, 82–83
  division, 91–93
  factoring, 107–112
  factoring difference, 107–108
  multiplication, 78
  multiply sum and difference of same terms, 81
  powers, 84–85
  squaring, 80
  squaring, pitfalls, 273–274
  squaring, radical equations, 180–181
braces, grouping and, 17
brackets [ ], grouping and, 17

• C •

Cartesian coordinates (graphs), 259
circles, radicals and, 57

coefficient
  adding/subtracting like terms, 67
  cubed binomials, 82–83
  synthetic division, 95–96
common denominator of fraction
  addition and, 32–33
  finding for, 30–32
  multiplication and, 35–36
  tips, 278
commutative law of addition and
    multiplication, 279
commutative property, 21–22
comparing numbers, number line, signed
    numbers, 7–10
comparisons in story problems
  age problems, 235–237
  answers, 242–244
  consecutive integers, 237–239
  work problems, 240–241
compound inequalities, 197
compounded interest, 214
consecutive integer story problems, 237–239
constants, binomial division, 91
converting
  decimals to fractions, 39–40
  fractions, 25–26
  fractions to decimals, 39–40
cubing
  binomials, 82–83
  differences, 108–109
  sum and difference, 83–84

• D •

decimals, fraction conversion, 39–40
denominator, fraction reduction, 103–104
Descartes' rule of sign, roots, 167–168
difference, cubes, 83–84, 108–109
distance formulas, 213
distance story problems, 226–228
distributing factors, expression
    multiplication, 77–78
distributive property, grouping symbols
    and, 19–20
distributive rule, expressions, 77–78
division
  binomials, 91–93
  cubes, 83–84
  exponents and, 48–49
  expressions, 68–69, 89–99
  fractions, 37–38

division *(continued)*
  monomials, 89–91
  polynomials, 94
  signed numbers, 13–14
  synthetic, 95–96
double root, quadratic equations and, 154

• E •

equality, equations, 131
equations
  absolute value, 177
  equality, 131
  linear (*See* linear equations)
  polynomial, 167
  quadratic, 153–166
  quadratic-like, 172–173
  radical, 177
equivalent fractions, 27–28
estimates, quadratic equation answers, 160–161
exponents
  answers to problems, 54–55
  changing radicals to, 60–61
  division and, 48–49
  expressions, simplifying, 62–63
  fractional exponents, 61–62
  multiplication, 47–48
  negative, 51–52
  raising to a power, 49–50
  scientific notation and, 52–53
expressions
  addition, 67
  answers to problems, 73–74, 86
  division, 68–69
  factoring, 101–106
  FOIL, 78–79
  multiplication, 68–69, 77–88
  order of operation, 69–72
  product of binomials, 107
  radicals, simplifying, 57
  rationalizing fractions, 59
  simplifying, division and, 89–99
  simplifying, exponents and, 62–63
  subtraction, 67
extraneous roots, radical equations, 177–179

• F •

factor/root theorem, polynomial equations, 170–171
factoring
  answers to problems, 111–112, 125–127
  grouping and, 120–121
  multiple ways, 109–110
  polynomial equations and, 171–172
  quadratic equation solutions, 154–157
  quadratic inequalities, 192–193

factoring binomials
  differences and sums of cubes, 108–109
  introduction, 107
  squares, difference, 107–108
factoring expressions
  GCF (greatest common factor), 102
  prime factorization, 101–102
  trinomials and, 107
factoring trinomials
  introduction, 113
  multiple ways, 118–119
factors, distributing one over many, 77–78
flipping proportions, 277
FOIL, 78–79. *See also* unFOIL method
formulas
  answers to problems, 216–220
  area formulas, 210–211
  distance formulas, 213
  introduction, 207
  linear equations and, 143–144
  perimeter formulas, 209–210
  Pythagorean theorem, 207
  quadratic formula, 157–159
  story problem answers, 229–234
  volume formulas, 211–212
fraction lines, grouping and, 17
fractional exponents, 61–62
fractions
  addition, 32–33
  answers to problems, 41–45
  common denominators, 30–32, 278
  converting, 25–26
  decimal conversion, 39–40
  divide by 2, 4, or 8, 279
  divide by 3 or 9, 278–279
  division, 37–38
  equivalent, 27–28
  grouping symbols similarity, 136
  improper, converting, 25–26
  linear equations and, 138–140
  mixed, converting, 25–26
  multiplication, 35–36
  multiplying through, 277–278
  percentages, 214–215
  pitfalls, 275
  proportions, 28–29
  rationalizing, 59
  reducing, 27–29, 103–104
  subtraction, 34
  tips, 277–278

• G •

GCF (greatest common factor), 102
  factoring, 107
  finding, 113–114
  quadratic equations and, 154

Index

trinomials, 113
trinomials, factoring, 118–119
geometry, story problems and, 224–226
graphing
  Cartesian coordinates, 259
  intercepts, 262
  lines, 262
  parabola, 267–268
  plotting, 261
  plotting points for lines, 261
  points, 259
  points, distance between, 266
  quadrants, 260–261
  slope-intercept form, 263–264
  slope of a line, 262–263
  slope of a line, perpendicular/parallel lines, 265
  transformations, 269–271
grouping, factoring by, 120–121
grouping symbols
  absolute value symbols, 17
  braces, 17
  brackets [ ], 17
  distributive property, 19–20
  fraction lines /, 17
  linear equations, 136–138
  parentheses ( ), 17
  radicals, 17

• H •

hypotenuse of triangle, 207

• I •

imaginary numbers, quadratic equations, radicals, 162
impossible answers, quadratic equations, 162
improper fractions, converting, 25–26
inequalities
  absolute value, solving, 195–197
  answers to problems, 198–203
  compound, 197
  introduction, 189
  linear, solving, 190–192
  negative numbers and, 189
  quadratic, 192–193
  solving by sections, 197
  statements, 189–190
integers
  consecutive integer story problems, 237–239
  definition, 237
intercepts (graphs)
  overview, 262
  parabolas, 268
  slope-intercept form, 263–264
interest formulas, percentages, 214–215
interrupted operations, 70
irrational numbers, description, 168–169

• L •

lead term, binomial division, 91
like terms, adding/subtracting, 67
linear equations
  addition/subtraction property, 131–132
  answers to problems, 145–151
  formulas, 143–144
  fractions and, 138–140
  grouping symbols and, 136–138
  introduction, 131
  multiple operations, 134–136
  multiplication/division property, 133–134
  proportions, 141–142
linear inequalities, solving, 190–192
lines (graphs)
  intercepts, 262
  plotting, 261

• M •

mixed fractions, converting, 25–26
mixtures, quantity story problems, 245–247
money story problems, 250–251
monomials, division, 89–91
MPZ (multiplication property of zero)
  polynomial equations solving with factoring, 171–172
  quadratic equations and, 154
multiplication
  binomials, 78
  cubes, 83–84
  exponents and, 47–48
  expressions, 68–69, 77–88
  fractions, 35–36
  signed numbers, 12–13
multiplication/division property, linear equations, 133–134
multiplication property of zero. *See* MPZ (multiplication property of zero)

• N •

negative exponents, 51–52
  pitfalls, 276
negative numbers
  absolute value and, 8–11
  distribution, pitfalls, 275
  inequalities and, 189
  signed numbers, division, 13–14
  signed numbers, multiplication, 12–13
number lines
  absolute value and, 8–11

comparing numbers, 7–8
  signed numbers and, 7–8
numbers, comparing using number line, 7–8
numerator, fraction reduction, 103–104

• O •

obtuse angles, story problems and geometry, 224
order of operation
  expressions, 69–72
  pitfalls, 274

• P •

parabola, plotting on graph, 267–268
parallel lines, slope and, 265
parentheses ( ), grouping and, 17
Pascal's Triangle, binomials, 84
percentages, 214–215, 248–249
perimeter, story problems and, 221–223
perimeter formulas, 209–210
perpendicular lines, slope and, 265
pitfalls
  distributing negatives, 275
  fractions, 275
  negative exponents, 276
  order of operation, 274
  radicals, 274
  squaring a binomial, 273–274
  squaring a power, 273
plotting graphs
  parabola intercepts, 268
  parabolas, 267–268
  points, 261
points (graphs), distance between, 266
polynomial equations
  answers to problems, 174–176
  factor/root theorem, 170–171
  factoring and, 171–172
  introduction, 167
  Rational root theorem, 168–169
  real roots, 167–168
polynomials
  division, 94
  tips, 280
positive numbers
  absolute value and, 8–9
  signed numbers, division, 13–14
  signed numbers, multiplication, 12–13
powers. *See also* polynomial equations
  answers to problems, 174–176
  binomials, 84–85
  quadratic-like, 172–173
  raising exponents, 49–50
  squaring, pitfalls, 273

prime factorization, 101–102
proportions
  flipping, 277
  fractions, 28–29
  linear equations and, 141–142
Pythagorean Theorem
  application, 207–208
  Phythagorus, 207

• Q •

quadrants, graphs, introduction, 260–261
quadratic equations
  answers to problems, 163–166
  double root, 154
  estimating answers, 160–161
  factoring and, 154–157
  impossible answers, 162
  introduction, 153
  MPZ (multiplication property of zero), 154
  quadratic formula, 157–159
  square root rule, 153–154
quadratic inequalities, solving, 192–193
quadratic-like equations, factoring, 172–173
quantity story problems
  answers to problems, 252–255
  mixtures, 245–247
  money problems, 250–251
  solutions, 248–249

• R •

radical equations
  answers to problems, 184–188
  description, 177
  square once, 177–179
  square twice, 180–181
radicals
  answers to problems, 64–65
  changing to exponents, 60–61
  expressions, simplifying, 57
  grouping and, 17
  negatives, quadratic equations and, 162
  pitfalls, 274
  quadratic equations, negatives, 162
  quadratic formula and, 157–159
  tips, 280
Rational root theorem, polynomial equations, 168–169
rationalizing fractions, 59
real roots, polynomials, 167–168
reciprocals, fraction division, 37–38
reducing fractions, 27–29, 103–104
reverse the sense, 276
roots

extraneous, radical equations, 177–179
real roots, polynomial equations, 167–168

• S •

scientific notation, writing numbers with, 52–53
signed numbers
  addition, 9–11
  answers to problems, 15–16
  division, 13–14
  multiplying, 12–13
  number line and, 7–8
  subtraction, 11–12
slides (graphs), 269–271
slope-intercept form (graphs), 263–264
slope of a line (graphing)
  overview, 262–263
  parallel lines, 265
  perpendicular lines, 265
solutions story problems (quantity/quality), 248–249
square root
  binomials, 80, 273–274
  binomials, factoring, 107–108
  double root, quadratic equations and, 154
  powers, 273
  quadratic equations and, 153–154
  radical equations, 177–179, 180–181
  rationalizing fractions and, 59
squares, radicals and, 57
story problems
  age problems, 235–237
  answers, 229–234
  area, 221–223
  comparing things in, 235–241
  consecutive integer problems, 237–239
  distance problems, 226–228
  geometry and, 224–226
  introduction, 221
  money problems, 250–251
  perimeter, 221–223
  quality, 245–251
  quantity, 245–251
  solutions problems, 248–249
  volume, 221–223
  work problems, 240–241
subtraction
  expressions, 67
  fractions, 34
  signed numbers, 11–12
sum, cubes, 83–84
supplementary angles, story problems and geometry, 224
symmetric property, tips, 280
synthetic division, 95–96

• T •

tips
  commutative law of addition and multiplication, 279
  fractions, common denominator, 278
  fractions, multiplying through, 277–278
  polynomials, 280
  proportion flipping, 277
  radicals, 280
  symmetric property, 280
transformations, graphing with, 269–271
translations (graphs), 269–271
transversal, geometry and story problems, 224
triangles, hypotenuse, 207
trinomials, 83–84, 107, 113, 115–117

• U–V •

unFOIL method, trinomials and, 115–117

vertex, parabola graphing, 267
volume, story problems and, 221–223
volume formulas, 211–212

• W–Z •

work story problems, 240–241

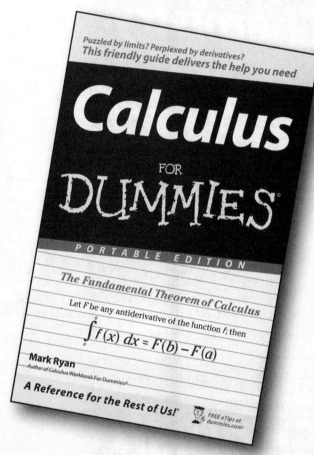

Also Available at Wal-Mart — For Dummies Portable Editions

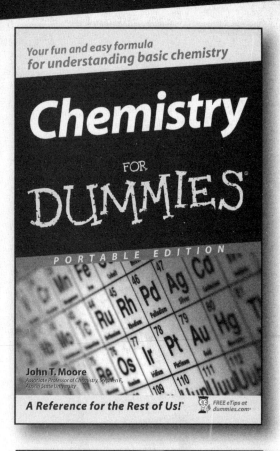

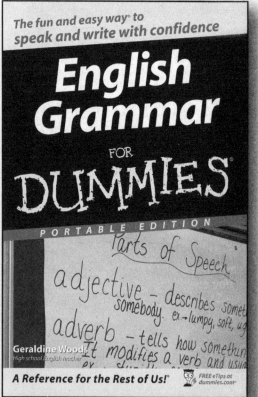

Also Available at Wal-Mart — For Dummies Portable Editions

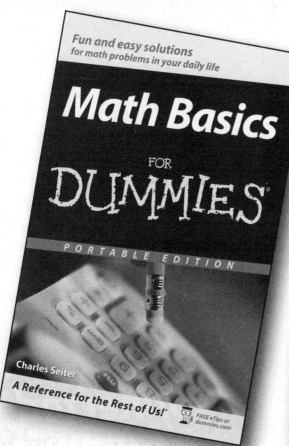

Also Available at Wal-Mart — For Dummies Portable Editions